TELEVISION

Recent Titles in
Greenwood Technographies

TELEVISION

THE LIFE STORY OF A TECHNOLOGY

Alexander B. Magoun

GREENWOOD TECHNOGRAPHIES

GREENWOOD PRESS
Westport, Connecticut • London

Library of Congress Cataloging-in-Publication Data

Magoun, Alexander B.
Television : the life story of a technology / Alexander B. Magoun.
     p. cm. − (Greenwood technographies, ISSN 1549-7321)
    Includes bibliographical references and index.
    ISBN−13: 978−0−313−33128−2 (alk. paper)
    ISBN−10: 0−313−33128−6 (alk. paper
    1. Television—History.   I. Title.
  TK6637.M34   2007
  621.388009−dc22   2007014283

British Library Cataloguing in Publication Data is available.

Library of Congress Catalog Card Number: 2007014283
ISBN-13: 978−0−313−33128−2
ISBN-10: 0−313−33128−6
ISSN: 1549−7321

First published in 2007

Greenwood Press, 88 Post Road West, Westport, CT 06881
An imprint of Greenwood Publishing Group, Inc.
www.greenwood.com

Printed in the United States of America

The paper used in this book complies with the
Permanent Paper Standard issued by the National
Information Standards Organization (Z39.48−1984).

10 9 8 7 6 5 4 3 2 1

Contents

Series Foreword

In today's world, technology plays an integral role in the daily lives of people of all ages. It affects where we live, how we work, how we interact with each other, what we aspire to accomplish. To help students and the general public better understand how technology and society interact, Greenwood has developed *Greenwood Technographies*, a series of short, accessible books that trace the histories of these technologies while documenting *how* these technologies have become so vital to our lives.

Each volume of the *Greenwood Technographies* series tells the biography or "life story" of a particularly important technology. Each "life story" traces the technology from its "ancestors" (or antecedent technologies), through its early years (either its invention or development) and its rise to prominence, to its final decline, obsolescence, or ubiquity. Just as a good biography combines an analysis of an individual's personal life with a description of the subject's impact on the broader world, each volume in the *Greenwood Technographies* series combines a discussion of technical developments with a description of the technology's effect on the broader fabric of society and culture—and vice versa. The technologies covered in the series run the gamut from those that have been around for centuries—firearms and the printed book, for example—to recent inventions that have rapidly taken over the modern world, such as electronics and the computer.

While the emphasis is on a factual discussion of the development of the technology, these books are also fun to read. The history of technology is full of fascinating tales that both entertain and illuminate. The authors—all experts in their fields—make the life story of technology come alive, while also providing readers with a profound understanding of the relationship of science, technology, and society.

Preface

Let me admit right now that I don't have a television and haven't watched one with any regularity since I left my parents' house and Sony Trinitron for college. Therefore opportunities to watch what billions of people take for granted are a treat, whether it is in the homes of friends and relations, Sarnoff Corporation's digital laboratories, on RCA's first color TV receiver from 1954, a Nipkow disc television from 1928, or on a Web site.

Second, I have no technical training beyond high school physics and college chemistry. Why, then, am I qualified to write on this subject? The answer lies in the David Sarnoff Library, which I began visiting while researching my dissertation in 1994. The nonprofit Library is housed within the David Sarnoff Research Center, formerly the RCA Laboratories and now owned by Sarnoff Corporation, which hired me to oversee the Library in 1998. That is a lot of Sarnoffs to keep track of, but the upshot is that the facility is the site of the invention of electronic, monochrome-compatible, color television and liquid crystal displays, as well as one of the world's centers for the innovation of digital television. Of necessity I have learned quite a bit about the history, business, and technology of television; the people who contributed to its development; and the people who collect, preserve, and promote that history today.

Seeing antique electronic equipment work adds immensely to one's re-spect for the people who invented it. Scott Marshall has restored numerous

TVs of the 1940s and 1950s. He explains what he does and what the earliest electronic television inventors did in their patents with equal clarity. Scott also introduced me to the wonderful people, experts, and collectors of all ages in the New Jersey Antique Radio Club. Maurice Schechter, chief engineer of DuArt Film & Video, has spent many years documenting and restoring RCA's World War II television systems. It is astonishing and chilling to see a complete chain of equipment, from camera to antennas to display and joystick, show what guided missile pilots worked with in the 1940s. Steve McVoy deserves kudos for organizing the Early Television Foundation and its stunning museum in Hilliard, Ohio. Steve's dedication to restorations is second to none, and life is incomplete without experiencing his Nipkow disc television. The annual conference of television collectors and researchers that he initiated in 2003 has enabled me to share my scholarship with an especially appreciative and critical international community, learn from other participants, and see a surfeit of obsolete television technologies in operation. In particular Ed Reitan has preserved more of CBS's color TV investment than anyone else and made it possible to watch color videotape recordings from the 1950s.

A book like this offers infinite opportunities to grossly simplify what occurs in the course of capturing a moving image in one place, transmitting it instantaneously, processing it, and recording or displaying it somewhere else. Those responsible for keeping me honest through their publications are listed in the bibliography. Many other video professionals also guided my writing. Among those I have spoken with in the past eight years, some have been especially patient in their explanations. These include Harold Borkan, Peter Burt, James Carnes, Joseph Castellano, Doug Dixon, Ernest Doerschuk, Rob Flory, Ray Hallows, John Holtzapple, R. Norman Hurst, Michael Isnardi, Harry Kihn, Bernard Lechner, Peter Levine, Dennis Matthies, Dennis McClary, the late Max Messner, Albert Morrell, Wendell Morrison, Edy Mozzi, Jeremy Pollack, Glenn Reitmeier, Martin Royce, Arthur Sarnoff, Edward Sarnoff, the late Alfred Schroeder, Fred Vannozzi, Gooitzen Van der Wal, John Van Raalte, Richard Webb, Larry Weber, the late Paul Weimer, Richard Williams, Niel Yocum, and Louis Zanoni. They have all contributed, not always for the purpose of this book, to my understanding of how television and its components work, in different technological formats and, as importantly, as a business. It is only fair to observe that most of these men are or have been associated with RCA, NBC, Sarnoff Corporation, or Thomson. It is also fair to note that they do not speak with one voice about David Sarnoff's electronic legacy, in its technical or commercial evolution. Whatever their perspectives, any interpretations or errors are mine, not theirs.

Historians are bound to their sources and experiences, and my immersion in the life and times of RCA over the past ten years invariably colors the narrative herein. Nonetheless, I believe I have written fairly about the roles of the various characters in the life of television as it ran its course in the United States. Moreover you can read it as a response to many histories of RCA produced since its demise in 1986, whose authors overreacted to the claims that RCA made for the sake of public relations. Exaggerated they may have been, but that is no reason to deny the reality of the claims themselves. Consider this work part of a third generation of television history, then, and if it provokes more scholarship on the subject, so much the better.

This is a work by a scholar, even if it is not purely scholarship—that is, including annotated documentation of the narrative's sources. Robert Friedel at the University of Maryland, College Park, is responsible for inspiring me to a career that lets me engage the process of technological change without having to do much math. Our associates in the Society for the History of Technology deserve credit for encouraging, pursuing, and synthesizing the research and writing that gives us all a better sense of where we've come from and how we got to where we are.

Finally, I thank Kevin Downing, my long-suffering editor, who conceived the series of which this book is a part. In waiting way too long for the manuscript, he defended the time it has taken me to write a life of television from the people who have a business to run to ensure that books like this make it to market.

I dedicate this book to my nieces and nephews—Brittany, Courtney, Kristen, Heather, Andie, Grace, Drew, and Ian—who will have to explain what television was to the next generation.

Introduction

This is a short history of television, a technological system that perhaps more people on the planet have engaged than any other, besides electrification, over the past fifty years. It focuses on television in the United States but refers to developments elsewhere as they contribute to the American system's growth. This book explains how television works, how we have changed our definition of what television is, how we arrived at the systems we have, and how we and other countries made choices along the way, for technological, commercial, political, or cultural reasons. Think of the variety of people involved: inventors, scientists, engineers, managers, executives, broadcasters, entrepreneurs, advertisers, regulators, performers, and consumers. They are the reason we can animate any technology and describe the trajectory of its development as a life.

This is a useful framework for understanding the evolution of a technology. That is, as long as we ride herd on the analogy and remember that television, the technology, does not do anything. People do things with and react to television in stages and relationships similar to those they themselves pass through. These stages include conception, birth, parenthood, school, work, and family; the latter refers to the offshoots of the mature working system, resulting in new systems and components that can be likened to cousins and descendants.

By conception we mean imagining a plausible way of accomplishing television, of sending moving images or video—instantaneously, in real time, live—from one place to another. Across the western world, scientists, engineers, amateurs, and students began imagining approaches to this technology in the late 1800s. This was a typical case of simultaneous conception; inventors followed, drawing on similar educations and inspiration by scientific discoveries and technological developments in the last quarter of the nineteenth century.

By birth we refer to the tricky question of credit for inventing a system that works: the reduction to practice of an idea. Who can demonstrate a working system that scans and captures a moving image, transmits those images over a distance—ideally without physical connection—and displays them on a receiver somewhere else? In the first quarter of the twentieth century, a variety of inventors did just that. These imaginative men, self-taught or university-trained, attracted supporters of their pursuit of something new and profitable, and designed and built electromechanical and electronic ways of making the concept of television work. Instead of one birth, we see many, and can assign a midwife's—or midhusband's—credit to four men.

Beyond invention lies the challenge of innovation, of parenthood and schooling. Who will shepherd an idea that works only as a demonstration in an attic or laboratory, under restricted and controlled conditions, to all the conditions of the marketplace and real world? As with many systems innovations, governments play a leading role, especially outside the United States. In the richest country in the world, however, even during the Great Depression of the 1930s, one large corporation sustained television's innovation and employed the people responsible for more research, engineering, marketing, manufacturing, and broadcast content than any other organization. Nonetheless, the U.S. government regulates usage of the atmosphere, and the electromagnetic waves radiating through it, on which all wireless transmissions rely. Further, beginning with World War II, its military branches started investing in the ability to see at a distance to keep its soldiers out of harm's way, or increase the precision of disarming the enemy.

At the war's end, television's commercial proprietors, or bosses, put the technology to work. A commercial system has to pay for itself, after all, and repay its parents or investors for their investment. The public and broadcasters in the United States bought literally into the notion of supplementing and supplanting radio, newspapers and magazines, and movies with their new sibling among the mass media. After a brief period of learning what the public liked that could be transferred from radio and what it preferred that was new, programmers settled into formats that reflected the interests, hopes, and fears of millions of viewers. In the United States this took place through the commercialization of programming over a few broadcast networks; elsewhere it usually meant government control of one or two channels. In the absence of home video alternatives, what did the chosen few broadcasters put on the air, what did viewers watch, and why?

Even as television appeared frozen commercially and technically for a generation, we overlook the fact that entrepreneurs and inventors developed television's relations. These include cable and satellite broadcasting, video recording technologies, and improved cameras and displays. Consumers began embracing these technologies with sustained commercial success in the mid-1970s, and their popularity has fragmented the market for people's eyes ever since.

By the late 1980s, almost 100 percent of the households in industrialized nations contained televisions using analog transmission standards. At the same time, increases in the availability of computer processing power, expanding cable and wireless networks, and refinements in cameras and displays pointed to the death of television as a discrete system of communication. The next generation of video technologies would be digital, not analog, in format, which increased radically the human ability to manipulate the signal. Digitization also merged video with other systems for transmitting electronically the ones and zeros of digital information. It would provide far more than high-definition imagery and sound rivaling that found in movie theatres, or offer more channels with more information to more specific markets or even households. With the adoption of digital standards at the end of the century, multinational corporations renewed the process pioneered in the 1930s of selling services and receivers to a dubious audience, while resourceful and imaginative individuals created new applications and industries for electronic communications, just as they had with radio and cable and satellite television. Combined with the riot of production options and delivery choices for live or personalized video, these twenty-first-century entrepreneurs finally realized the cartoonists' predictions of the nineteenth century for the many possible uses of seeing at a distance.

If you read this history for reference purposes, you might pause to think about a few themes. One is the cause of technological improvement. Perhaps you have seen earlier generations of television receivers, on the Internet or at museums. Or you have lived through successive changes in the technology. In both cases, there is a good chance that you will remark or think at how far we have come in terms of picture size, resolution, versatility, price, content choice, etc. This constant drive to add value to a product is a fundamental part of western culture and its capitalist market economy. We take such technological improvements for granted; yet it is worth asking who is responsible and who pays for them, if only to understand a culture that originated in northwestern Europe 500 years ago and has spread, in part through television, over the rest of the globe.

Can we blame or praise the inventors—those tinkerers, scientists, engineers, and their assistants—for seeking to build the video equivalent of a better mousetrap? Or should we look at the entrepreneurs—investors, businesspeople, military agencies, corporate executives, bankers, directors, writers, actors—who make continued research and development possible? Or shall we assign responsibility to the consumers—the government regulators, armed forces, radio station owners, early adopters of the latest improvements, and the 80 percent of a population that represents a middle-class mass market? While only the inventors test and know the art of the possible in the domains of physics and chemistry that make television work, the entrepreneurs have to set priorities among the possible options to improvement, whether to improve at all, or whether to invest in some other means of profit instead. Consumers represent the ultimate arbiters in ways similar to producers but with more finality: is this improvement, this upgrade, worth its price in time and money? Will it improve the quality of their lives and leisure by substituting for some current part of their habits and schedules?

Second, for technological change to take place, those responsible for it need some motivation to justify their lives spent in this field. You will see that inventors and entrepreneurs are not in it only for the money. We all need to make a living, but there are less stressful alternatives than creating something new, under deadline and to a budget, to be tested in the unpredictable marketplace. The desire for an income, much less wealth, is almost always tied up with a variety of other inspirations and incentives. We usually associate creativity with the arts—literature, music, painting, and sculpture—activities designed by the artist to elicit an emotional response by an audience. We do not usually connect creativity to the objects of the material world and their innovators. Consider the imagination, however, in designing an electronic circuit to cancel noise in a broadcast signal, or of a chemist to fabricate a phosphor that rises to and recedes from true red in just the time it takes for one frame of video to flash on a picture tube. These creations do not elicit awe or joy until we see them in retrospect or in an educational setting, until we see what past inventors did with what now appear to be primitive tools and knowledge. Yet engineers and scientists lead creative lives, trying to define what makes something "better" and then building and using tools to accomplish that more quickly and effectively than other people working toward the same goal. This is a competitive environment, but one only more explicit than that surrounding the artist, who struggles for attention, sponsorship, and investment as well.

Third, the life of television over the past 130 years offers a neat reflection of the health of American ingenuity and power relative to the rest of the world. Initial efforts to invent television originated in Europe in the nineteenth and early twentieth centuries before the United States asserted its economic power and became the center for its innovation and diffusion in the middle third of the 1900s. In the 1960s, Japan began an Asian revolution in broadcast and consumer electronics innovations that destroyed an American industry. At the beginning of the twenty-first century, the American economy relied heavily on the sale of services and growing consumer indebtedness while China, Japan, South Korea, and Taiwan controlled the design and manufacture of displays and cameras, along with the manufacture of other electronic devices. However one connects these phenomena and whatever their consequences for the technical, commercial, and political influence of the United States, few American observers regarded the trends as positive ones.

This book poses two questions without dwelling on their answers at length: who invented television, and to a lesser extent how television's content relates to the larger culture. The issue of invention is one of definition, and here we will touch on the technical, legal, and other factors that shape definitions and complicate the notion of "first." As for the issue of technology and culture, programs reflect the changing values of the society in which they are broadcast, filtered through the owners and operators of television stations and networks. For all of those distortions, and the lack of specifics as to why individuals watch a particular program, we gain some sense of the concerns and interests, voiced and unspoken, of a particular place and time. With the content of television we have a useful counter to the often unspoken assumption that a materially wealthy society enjoys cultural superiority over other, less technologically advanced societies, or its own ancestors. In other words, for all of the remarkable technological developments described here, beware of concluding that the quality of human nature has improved alongside the inventions that raise the quality of material life.

Timeline

1873   British discovery of photoelectric effect in selenium sparks international efforts to build television cameras and displays.

1884   German Paul Nipkow applies patent on electromechanical television systems using perforated, spinning, scanning and display discs that took on his name.

1907   Frenchman Constantin Perskyi coins the term "television."

1908   Briton Alan Archibald Campbell Swinton proposes electronic television system.

1912   Russian Boris Rosing uses cathode-ray tube (CRT) for electronic television display.

1924   Briton John Logie Baird demonstrates first television system, using Nipkow discs.

1925   Vladimir Zworykin demonstrates electronic television camera and display to Westinghouse Corporation executives.

1928   Philo Farnsworth demonstrates electronic television system including wireless transmission.

1939   David Sarnoff introduces NBC's regular television broadcast schedule using Radio Corporation of America (RCA) system.

1941   Federal Communications Commission (FCC) approves electronic television standard based on RCA system.

1948   Peter Walsonavich installs first cable television system in Mahanoy City, Pennsylvania.

1950   RCA demonstrates electronic color television system using single cathode-ray tube display.

1952   NBC reporters use handheld RCA television cameras at presidential conventions.

1953   FCC approves color television standard based on RCA system.

1956   Ampex introduces first commercial television tape-recording system.

1962   AT&T transmits television signals across Atlantic Ocean by Telstar satellite.

1966   Donald Bittzer and H. Eugene Soltow demonstrate first plasma display.

1968   RCA demonstrates first liquid-crystal displays (LCDs).

1971   Bell Laboratories' Willard Boyle and George Smith demonstrate first camera using a solid-state charge-coupled device (CCD).

1975   Briton Steven Birkill receives satellite television on homemade dish antenna.

1977   JVC and Matsushita begin selling VHS videocassette player/recorders.

1984   Epson begins selling color, LCD, wristwatch televisions.

1985   Sony begins selling CCD camcorder.

1988   Sharp Electronics introduces first 14-inch LCD color TV.

1993   Briton Tim Berners-Lee invents World Wide Web system for the Internet.

1994   ABC puts the first broadcast network program on the Internet.

1994   Thomson begins selling home, digital, satellite, television system.

1996   FCC approves digital and high-definition television standards developed by multinational consortium.

1997   Multinational consortium begins selling digital videodiscs (DVDs).

2006   Thomson and Sony close color CRT research facilities.

2007   Apple announces iPhone that combines cellphone with camera, LCD, keyboard, and Internet access.

1

Conception, 1873–1911

"If I have seen farther, it is by standing on the shoulders of giants." One scholar traced this phrase to the ancient Greeks, some 2,500 years ago. It has a double meaning for all of us who watch television, a technological system based on the cumulative contributions of thousands of people. These include workers on technologies predating, influencing, and contributing to television. For, before someone could transmit moving pictures instantaneously in a practical manner, he needed the tools to offer the prospect of fulfillment. A new technology like television, one that makes a significant leap in human control of the environment, first requires advances in a variety of sciences and technologies to make thinking about its invention more than a daydream.

By the 1870s, enough of these advances had occurred in transmitting data and sound so that researchers and inventors across Europe and the northeastern United States became interested in taking the next step, of transmitting moving pictures instantly. Looking back, this seems like fantasy. At the time, large batteries provided the only practical source of electricity; the subatomic electrons that constitute the basis of electrical energy would not be identified for another twenty years; and the wealthiest people illuminated their homes with light from burning jets of gas. There were no sound recordings, no motion pictures, no telephones, no radios, no computers. The term "television" awaited the new century and no one agreed on what the new technique and technology should be called. Instead all interested parties identified them in terms of familiar systems.

These men—and they were all men—shared centuries-old European traditions of scientific research and invention. The significant spur to their interest was the development of the electrical technologies of cable telegraphy, facsimile, and telephony between 1837 and 1876. These provided media for the instantaneous transmission of signals, still pictures, and sound through insulated copper wires. Other stimuli included numerous moving-image technologies and the refinement of electric light bulbs and light systems in England and the United States in the 1870s. Alexander Bell's telephone added sound to the dots and dashes of the telegraph; light bulbs suggested a way to represent each cell, or pixel, in a video display. Along with Thomas Edison's invention of sound recording and reproduction, these highly publicized technologies inspired the thoughts, hopes, and convictions of scientists, engineers, inventors, and cartoonists that television might be more than a fantasy. In December 1878, a British cartoonist's "telephonoscope" suggested that Edison would soon enable parents to see and speak with their children on the opposite end of the world. Adding sight to sound appeared to be a natural progression, technically and culturally.

This new system would transmit by wire, because the demonstration of electromagnetic waves and the prospect of radio signal transmission on those waves still lay ten years in the future. The conceivers also built on the earlier electrical media as well as a variety of scientific discoveries. During the nineteenth century men across Europe began explaining the mathematical properties and relationships of atoms to energy in its electrical, magnetic, and light forms. They shared this information with other scientists, inventors, and engineers. Technical meetings, postal systems, and amateur and professional journals all helped spread information on developments in engineering and physics. Contributors wrote or spoke in English, French, or German, the languages of the great powers of Europe and North America, and they relied on knowledge of additional languages or translations to learn what others were doing to advance, directly or indirectly, the concept of television.

TELEVISION'S INGREDIENTS

Television requires a number of basic techniques and technologies to make it plausible. These include a material or device sensitive enough to capture the light reflected from an object over time and convert, or transduce, all its changing brightness and contrast, if not color as well, into electricity. More

PUNCH'S ALMANACK FOR 1879.

EDISON'S TELEPHONOSCOPE (TRANSMITS LIGHT AS WELL AS SOUND).

(*Every evening, before going to bed, Pater- and Materfamilias set up an electric camera-obscura over their bedroom mantel-piece, and gladden their eyes with the sight of their Children at the Antipodes, and converse gaily with them through the wire.*)

Paterfamilias (in *Wilton Place*). "BEATRICE, COME CLOSER, I WANT TO WHISPER." *Beatrice (from Ceylon).* "YES, PAPA DEAR."
Paterfamilias. "WHO IS THAT CHARMING YOUNG LADY PLAYING ON CHARLIE'S SIDE?"
Beatrice. "SHE'S JUST COME OVER FROM ENGLAND, PAPA. I'LL INTRODUCE YOU TO HER AS SOON AS THE GAME'S OVER?"

With publicity spreading about Alexander Bell's telephone, Thomas Edison's phonograph, and proposals for seeing at a distance electrically, some farsighted observers suggested that *television*, a term coined only in 1900, was not far off. Here an English cartoonist in 1879 imagined an interactive, high-definition, flat-panel display that would enable families divided by imperial duties to stay in touch. David Sarnoff Library

specifically it has to convert the photons of light into equivalent numbers of electrons. Because the efficiency of transduction is never 100 percent, the series of electronic images has to be amplified above the electronic noise in the system as well as transmitted to a receiver. The receiver has to include some means of converting or transducing the electronic images back to gradations of light, at least from white to black, again if not the color spectrum of visible light.

From these basic requirements—transduction from light to electrons, amplification, transmission, transduction from electrons to light—follow several others. The object has to be scanned: light from any number of points on it is converted to electricity and sent to the receiver. It is usually impractical to provide a channel of transmission for each point. Therefore, reducing a two-dimensional image to a one-dimensional transmission means the light of the image has to be scanned and transduced in some linear fashion. Then it can be sent as a continuous electromagnetic wave to be reconstructed at the receiving end. The time to scan each successive image requires a pause between scans. So the scanning has to occur quickly enough to exploit the perceptual trick our eyes and brains play in retaining the light from an image for a fraction of a second. This persistence of vision permits the illusion of continuity, and eliminates the flicker of a moving object between successive scans. Sequential scanning of an image, and reproduction of those scans at the receiver, requires synchronization: between the capture of the lines of a frame of an image and the display of the lines of that same frame. Finally, in the course of conversion, amplification, and transmission, the image loses some information and gains noise—undesired or distorted electronic signals. Some sort of selective amplifier is necessary to make the reproduction as close to the original as possible.

Individuals had already addressed some of these issues by the second half of the nineteenth century. Although Ptolemy observed the persistence of vision in ancient Egypt, no one thought the phenomenon especially useful until the early 1800s. Various mechanical devices and toys, starting with the magic lantern in 1671, enabled the display and projection of moving images well before the development of photography and motion pictures. If lecturers or entertainers project a series of images from glass slides on the wall of a room within a given time, at some point the series appears continuous. With the acquisition of successive images, the viewer's brain superimposes movement from one frame to the next, giving a viewer the impression of continuous movement.

But how could one turn the light reflected off a succession of images into an electrical signal and then reverse the process instantaneously? Although there was nothing simple about making a suitable display, among the components of television technology, image capture represented the singular challenge. The problem gave rise to two approaches. One was electrical and then electronic, and used no moving parts. The other was electromechanical, involving some moving device to collect light at a series of points on the image. And while we have traditionally considered the electronic approach more modern, it having been the standard method for the last half of the twentieth century, keep in mind that microelectromechanical switches and spinning color wheels are now part of the equipment for most projected video displays.

ELECTRICAL CONCEPTIONS OF TELEVISION, 1873–1881

In the nineteenth century, electrical conceptions of television arose first, based on the discovery that the element selenium is photoconductive. That is, it conducts electrical charge in relation to the amount of light shone on it. This was not a serendipitous operation related to a search for electrifying images. Instead, in 1872 Briton Joseph May discovered accidentally that selenium rods used in the Atlantic Ocean telegraph cable lost their resistance to electrical conductivity when exposed to sunlight. His supervisor, Willoughby Smith, reported this phenomenon in a letter read at the Society for Telegraph Operators' annual meeting in London the following February. The only immediate reaction was the chairman's, who thought that "they would hear a good deal in the future" about selenium's photoconductivity as the basis for "a perfect photometer" (Fisher and Fisher, 1996, 10–11).

May and Smith had more ambitious ideas. Smith proposed a "visual telegraph," using an array of selenium photocells that conducted electricity in proportion to the light reflected from an image. The variable amounts of electricity would then be reconstituted in another array of undefined elements at a receiver some distance away. Each cell in what we would call a camera would be wired to another cell in the display. May tried to build a selenium cell-based camera, but Lieutenant R. E. Sale of the Royal Engineers explained to England's Royal Society why this effort was doomed to failure. The element reacts almost instantly when exposed to light, but it only slowly loses its photoconductivity when the light is removed. Therefore the signal from one image, or frame, would overlap on the display with those following it. Nor is selenium very efficient as a transducer. The amount of electrical current generated by light is insufficient for a display of practical brightness. Together these inherent drawbacks make it almost impossible to obtain the persistence of vision effect when a selenium array reacts to a series of images.

Despite Sale's findings, the prospect of instantly transmitting moving images was a powerful attraction to researchers at the cutting edge of science and technology. Word of the discovery spread through technical and popular publications and other amateur and professional researchers tried to succeed where Smith and May failed. At the time, they lived in a world where scientists across Europe seemed to be well on the way to understanding and uniting the relationships between the phenomena of light, heat, magnetism, and electricity. In the United States, Alexander Graham Bell and Thomas Edison led a variety of self-taught inventors in accomplishing what professional authorities never imagined or insisted was impossible. Bell's telephone and Edison's phonograph, accomplished in 1876 and 1877, stimulated research in the implausible. Simply because someone presented a paper to one of the world's most prestigious scientific organizations, denying the practicality of selenium, was no reason not to look for another approach.

Thus Portuguese physics professor Adriano de Paiva wrote an article in 1878 and published a booklet describing an "electric telescope" in 1880. Writing in French rather than Portuguese to increase its circulation, he described a camera plate covered with selenium. In 1878 Bell also proposed development of what he called his most important invention, the "photophone." Over the next three years this provoked more theories, underfunded experiments, and publications because he described the use of light and selenium in a way that some readers interpreted as making them capable of transmitting images as well as sound.

Among them was Constantin Senlecq, a French lawyer, who described a "telectroscope" early in 1879. Here a spring-mounted selenium photocell would move over light from an image focused on a glass lens. The cell would transmit varying amounts of electricity to an electromagnetically driven pencil, which would illustrate the image on a receiver in accordance with the amount of light in the image. Senlecq's report on this new form of facsimile appeared in the *English Mechanic* and was followed a month later by Dubliner Denis Redmond's description of experiments with another "electric telescope." Redmond explicitly modeled his system, which he built, on the human eye, hoping to transmit ten images a second to exploit the persistence of vision effect "just as a rapidly whirled stick forms a circle of fire" (Fisher and Fisher, 1996, 12). His imager used photocells transmitting through wires to a similar array of platinum filaments, which would glow in proportion to the light reflected on each cell. This proposal, like many others, required an electrical wire or circuit for each cell to each light in the display, which ultimately limited the resolution of the image for practical reasons.

Others joined in. Carlo Perosino of Italy suggested in 1879 that a selenium-tipped stylus scan an image on a metal surface and a platinum point, synchronized with the scan, reproduce the image electrochemically on a receiver. W. E. Ayrton and John Perry reported to *Nature* in April 1880 on a proposed system using a mosaic of selenium cells. Each of these was to be connected by insulated wire to an arrangement of electromagnetically operated shutters. In the June issue of *Scientific American* that year, George Carey in Boston, Massachusetts, described a receiver using an array of carbon or platinum filaments in an evacuated glass tube, each glowing in relation to the electricity received from a matching cell in a "selenium camera." Having been the first to call a video imager a camera, Carey claimed this system was "ready to go into the stores" (Abramson, 1987, 11; Fisher and Fisher, 1996, 13).

That November of 1880, Frenchman Maurice LeBlanc became the first person to propose scanning an image onto a single photocell in an article in *La Lumière Electrique*. Two mirrors, one scanning horizontally, the other vertically, would deflect light from an image onto a photocell. The current induced would flow to a synchronized light valve using another pair of mirrors to modulate, or change, the amount of light emitted and displayed at a receiver. Across the English Channel, William Lucas optimistically followed with a similar scanning approach to moving images eighteen months later. For the display, he proposed using two Nicol prisms, made of crystalline calcium carbonate and set at right angles to one another. These would focus and move vertically and horizontally a light beam transmitted from a selenium scanning element. An electromagnet also reacting to the current generated by the selenium scanner would control the intensity of the light, ensuring that "an image in light and shade will be formed upon a screen at the receiving end, an exact counterpart of that at the transmitting end; and, more than this, every exact change in the image in the transmitter will be faithfully depicted upon the screen of the receiver." Each scan would take place in time to exploit the persistence of vision—in theory, anyway, for Lucas never demonstrated his concept (Udelson, 1982, 16).

At Great Britain's Physical Society in the spring of 1881, two sets of inventors showed how selenium could work for a single image. Shelford Bidwell scanned, transmitted, and reproduced "simple designs in black and white . . . a butterfly with well defined marks upon its wings, and a rude drawing, in broad lines of a human face" (Burns, 1998, 57). The photocopying process took about twenty minutes to develop the images, each about two inches square. Ayrton and Perry followed their earlier proposal with a demonstration of how changes in light shone on one cell of selenium could be projected on a screen. This too was a time-intensive process since the inventors did not change the amount of light too quickly. As for the prospect of building a camera using multiple cells, each wired to a counterpart in the receiver, Ayrton and Perry could only suggest that the new telegraph technology of multiplexing, sending several message signals on one line, might be a solution.

These demonstrations and attendant publicity in *Nature* marked the high point of selenium scanning for television. Perry echoed the cartoonist's earlier commentary and predicted that "the people of one hundred years hence . . . will probably be able to see one another's actions at great distances, just as if they were close together" (Burns, 1998, 58). But it would not be through the poorly understood operation of selenium photocells. The demonstrations also showed the impracticality of using the element for anything more.

All of these proposals represent concepts with a common goal. They also share a lack of investment beyond the time and money expended by the inventors themselves. No one reported reliably a demonstration of instantaneous electric transmission of a single image, much less a series. Still, we should remember the principle of photoconductive transduction, for later scientists would apply it to more sophisticated compounds. Julius Elster, Hans Geitel, and Alexander Stoletow began this process when they demonstrated the photoelectric activity of alkali metal amalgams in 1890.

NIPKOW'S MECHANICAL TELEVISION, 1884

The primacy of practical conception goes to a German graduate student, Paul Nipkow. Living in a rented room that doubled as his laboratory and workshop in Berlin, he received in 1883 the loan of a Bell telephone. After building his own microphone and transmitting words to a neighbor's attic, he began to consider the challenge of images. That Christmas Eve, Nipkow hit upon the solution. Despite the holidays and difficulties of hiring a patent lawyer, by January 6, 1884, he had filed for what became German patent number 30,105. Here was the solution to effective scanning and display, even though the young inventor drew, as his elders did, on selenium for the transduction between light and electricity. Nipkow's special contribution was a pair of discs for a camera and receiver. The Nipkow discs, as they became known, featured small 24 rectangular holes or apertures arranged in a spiral formation near the rim. Each hole, as it spun past the light reflected by or transmitted through an image, scanned a line, slightly curved by the arc of the disc's rotation. One rotation of the disc generated a 24-line frame. The frequency of rotation determined the number of frames per second.

The light passing through the spinning holes hit a selenium photocell that transmitted a current varied by the amount of light on the cell. For the display, the current created an electromagnetic field placed between the two Nicol prisms first suggested by William Lucas in front of a separate light source. Drawing on Michael Faraday's mid-century discovery of a magneto-optic effect, Nipkow fixed the prisms at right angles and let the electromagnetic field modulate the light from the source in proportion to the current emitted by the selenium cell. Another Nipkow disc, synchronized to the spin of the first by the vibrations of tuning forks, enabled the reconstruction of the images as continuous frames using the persistence of vision. Like most of the other inventors, Nipkow never built his *elektrisches Telescop*. His discs, however, became the method on which the first television systems were based some forty years later, in the 1920s.

THE TROUBLE WITH TRANSDUCTION, 1884-1910

What took so long? Inventors remained stuck on the problem of transduction. Even more proposals followed Nipkow's over the next twenty years. The vast majority proved fruitless for later development, based as they were on selenium for converting light to electricity. In 1889 Jean Lazare Weiller suggested scanning an image by outfitting the outside of a revolving drum with a series of mirrors. Each mirror would be tilted at an angle slightly different from those before and after; the light from an image would reflect in a series of lines from the mirrors onto a selenium cell. Various inventors and companies used mirror drums in the 1920s and early 1930s for low-resolution imaging; they did not, however, draw on Weiller's display. This transduced the electric current from the photocell into a telephone diaphragm, where the sound waves vibrated a gas jet. The variations in light from the jet reflected off another mirror drum onto a screen. Another inventor and apparent fraud, Jan Szczepanik, offered a possible solution to the photocell problem by rotating a disc of selenium before the light from an image. This would theoretically allow the element to refresh its capacity for transduction. Szczepanik promoted his technology vigorously in the late 1890s, speaking of its ability to transmit moving images in color at great distances and of a syndicate of French investors willing to support the exhibition of the technology at the Paris Exposition of 1900.

The Exposition took place, but Szczepanik was never heard from again. That August, Constantin Perskyi introduced the term "television" to the world through a paper with that title, which he read at the International Electricity Congress. It took eight years for the British patent office to include the term in its subject classifications, and another two years for it to appear in an American newspaper.

By one count, inventors made eighteen new proposals for television systems in the first decade of the twentieth century. Virtually all of them were based on selenium photocells and mirrors or Nipkow discs. At this point, television might be consigned to conception only, its proponents lacking the tools to make it a reality. But two of the proposals drew on innovations in another field, and here we see how other, apparently unrelated, developments shape the future of an unborn technology.

CATHODE-RAY TUBES, 1897-1911

The evolution of research in the behavior of electricity in a vacuum began 150 years earlier in Germany and Great Britain. The invention of Leyden jars to store and discharge an electric charge led to nearly a century of experiments on the behavior of electricity in gas-filled or vacuum tubes. Other researchers improved the quality of pumps and evacuation techniques. In the 1850s, German Heinrich Geissler significantly improved the vacuum pressure in glass tubes and Professor Julius Plücker found that a magnet placed next to a Geissler tube containing two electrodes aligned the lines of electric charge between them. By the 1870s, Scotsman James Clerk Maxwell had announced his theory relating electricity and magnetism, and other scientists confirmed that the "rays" emanating from a negatively charged electrode, or cathode, were negatively charged particles. Maxwell's notion of a "molecule of electricity" gained substance over the next twenty-five years until England's Joseph J. Thomson announced in 1897 his measurements of the charge, mass, and velocity of an electron. To carry out four years of experiments Thomson invented a cathode-ray tube (CRT) that used electric or magnetic fields to deflect the electrons emitted from a cathode before striking phosphors deposited on the other end of the tube. The same year that Thomson announced his findings, Ferdinand Braun began producing a commercial version of the CRT for laboratory research on electric and early radio circuits.

It took nine more years of experimenting and tinkering before Max Dieckmann and Gustav Glage in Germany proposed using a CRT not just for tracing electron behavior in a circuit but, in the words of their 1906 patent application, "the transmission of written materials and line drawings" (Burns, 1998, 115). One of these drawings depicted a beer stein. The Germans generated the displays in nearly real time, but the images displayed on the face of their tube were not scanned. Instead they had attached a pencil to sliders on x- and y-axes, each wired to vary in electrical resistance with the movement of the pencil. The varying resistance in two dimensions was transmitted to two sets of electromagnetic coils around the CRT. The coils deflected the electron beam inside the tube and its illumination of the phosphors on the display end of the tube. It was not television but it was an electronic display, about one and a quarter inch square. Dieckmann, who would continue to explore electronic television into the 1920s, noted that "the cathode ray is well worth the attention of inventors in search of apparatus destitute of inertia" (Burns, 1998, 116).

Less than a year later, in July 1907, at the Technological Institute in Saint Petersburg, Russia, physicist Boris Rosing took this idea further. He applied for a patent on a hybrid television system, one that combined a camera using two sets of rotating mirrors to scan an image onto a light cell, and a CRT for the reconstituted display, to improve on earlier "insufficiently mobile and sensitive" receiving devices. Over the next four years, aided by his student Vladimir Zworykin, Rosing tried to make his system work. By May 1911 he could report the scan and display of "four luminous bands" (Abramson, 1987, 36).

Rosing resorted to a selenium photocell as the source of the image, after failed experiments with the alkali amalgams with which Elster and Geitel had been experimenting for twenty-four years in Germany. Rosing also found that synchronizing the image scanner to the CRT display became a challenge. His first CRTs processed the electron beam in two stages. First, the beam was deflected away from an aperture in the tube in relation to the brightness of the image. The electrons that continued through the aperture were then deflected electromagnetically to scan the phosphor screen. This process gave a blurry image, leading Rosing to propose linking the velocity of the beam to the gray scale of the image. The beam would be moved quickly in relation to lighter portions of the scanned image, and slowly over darker parts.

Rosing could not make this work because of problems with synchronizing the speed of the electronic display with the slower movements of his electromechanical camera. More intensive technical developments were needed before an all-electronic system could be demonstrated, but the Russian's work attracted notice in Europe and the United States. In 1910 he and another writer predicted that television would "permit man not only to commune with other human beings, but also with nature itself." They proposed that an "electric eye, a help to man in peace, will accompany the soldier and facilitate communication between all members of society." Rosing imagined its use in industrial exploration of the oceans, in mining, and as a remote sensor "on lighthouses and at guard posts" (Burns, 1998, 121). As with so many scientists of this period, blind to the prospect of a consumer culture, he hardly imagined entertainment as a useful or attractive application. But the idea that new uses of natural laws of physics and new technologies would improve life for humanity drove Rosing and his protégé Zworykin throughout their careers.

CATHODE-RAY CAMERAS, 1908–1920

While the Germans and Russians began experimenting with electronic displays, others in Europe sought ways to overcome the problem of scanning a moving image. Twenty-seven years after he demonstrated a form of facsimile, Shelford Bidwell reappeared in the letters column of *Nature* to rebut the blithe assertions of a Frenchman that "within a year . . . we shall be watching one another across hundreds of miles apart" (Abramson, 1987, 28). By 1908

Bidwell had spent too much time on the problems of transduction to accept this on faith alone. In his letter he explained the challenges of synchronizing the operation of a camera and receiver with enough bits of video information to satisfy even viewers of a half-tone newspaper image. An equivalent video picture about two inches square would require 16,000 bits or pixels, refreshed at least ten times a second to show movement effectively. Such a transmission frequency, 160 kilohertz, seemed remote indeed when radio researchers were just beginning to transmit sound with bandwidths almost one hundred times smaller.

Instead Bidwell proposed an even more impractical approach: connect the camera and receiver physically by a cable eight to ten inches thick. This would carry signals from each of 90,000 selenium cells 100 miles away to a display two inches square. The cost? A little over $6,000,000, and triple that for color. Astonishingly enough, several researchers pursued this impractical approach, while a self-taught Scotsman rejected Bidwell's solution in favor of the first all-electronic method of television.

Alan Archibald Campbell Swinton was a consulting electrical engineer who started experimenting with a CRT that he had ordered from Ferdinand Braun around 1900. Three to four years later he mounted a selenium plate inside a CRT in an attempt to display changes in current flowing through the circuit. When Bidwell's letter appeared on June 4, 1908, Swinton responded with a letter published two weeks later. All the problems of scanning vast amounts of data could be resolved by synchronizing two CRTs and their associated electromagnetic deflection circuits. Displaying an image was not the challenge; the problem remained in a fast-acting transducer for the CRT camera: "Possibly no photoelectric phenomenon at present known will provide what is required in this respect, but should something suitable be discovered, distant electric vision will, I think, come within the region of possibility" (Abramson, 1987, 29).

Three years later, the Röntgen Society of London, organized for the study of X-rays, elected Swinton its president. In his address to the society at its annual meeting on the evening of November 7, 1911, Swinton expanded on his earlier suggestion to the problem of "Distant Electric Vision" and presented a possible solution for the electronic camera (Abramson, 1987, 38). In this new tube filled with a positively charged or ionized gas, like sodium, the light from an image would be projected through a metallic mesh screen onto a matrix of tiny metal cubes made of rubidium. These cubes were insulated from one another and charged by an electron beam scanning them on the other side of the matrix. The beam would give all the cubes a negative charge; where light from the image stimulated the rubidium, electrons would discharge and conduct themselves through the sodium vapor to the mesh. There the signal, proportional to the amount of light on each of the cubes, would be conducted to metal plates in the CRT display. The amount of charge would deflect the electron beam proportionately from an aperture, as in Rosing's patent, and the remaining electrons in the beam would be electromagnetically deflected horizontally and vertically across the phosphor screen to reproduce the image. To replicate the resolution of photoengraving, Swinton proposed scanning 200 vertical lines per frame, refreshed ten times a second in keeping with what was known about the persistence of vision.

Swinton conceded that the problem of photoelectric "sluggishness" remained, but rationalized that it "in no way interferes with the correct transmission and reproduction of the image, provided all portions of the image are at rest.... In fact, sluggishness will only cause changes in the image to appear gradually instead of simultaneously" (Burns, 1998, 125). Perhaps he was counting on the relatively low refresh rate to give the photoelectric response time to recover for each scan. How many in Swinton's audience took notes on his after-dinner speech is unknown, but *The Times* of London reported it and the society reprinted his address the following January. Given the primitive nature of the science and technology of electron tube physics, electronic circuits, vacuum technology, and photoelectric chemistry, Swinton's elegant ideas seemed a distant vision indeed. But they and Nipkow's discs proved to be the conceptions on which others reduced paper inventions of television to practice.

2

Birth of a Technology; or Invention, 1912–1928

◆

DEFINING INVENTION

This chapter will not tell you who invented television—or, rather, the television system that you may watch today. Instead it focuses on the stories of four men often assigned that honor and explains how each of them invented a television system in the 1920s. The difference between the definite and indefinite articles "the" and "a" is crucial in defining invention, a distinction that makes the challenge more manageable. In the last chapter we found men proposing and attempting two approaches to television, one electromechanical in design, the other purely electronic, each with the components that comprise a system. Does it matter? To those obsessed with assigning places, yes, but as scholars show so often and participants know too well, crediting one individual for an invention, however one defines it, amounts to an exercise in futility as the technology becomes more complex.

The birth of the four systems took place almost simultaneously in the 1920s. Most popular historians describe this process as a race. In an abstract way, inventing rival approaches for the same product is. All involved have access to knowledge about the state of the art, similar technological components with which to make the new system work, and commercial pressure to attain legal priority. Ultimately inventors want to see their babies become commercially useful, if not profitable. But their strategies for accomplishing this goal vary hugely.

We can lump the creators in two categories: lone inventors and corporate inventors. Many of us applaud lone inventors as eccentric rebels who think of unconventional approaches to help spur revolutionary change in a field, but some of them simply prefer working independently. They have difficulty fitting into the large organizations that structure modern society because of the nature of their egos or their education. They are more suited to leading a small business in whatever direction their passion takes them and their followers. Lone inventors hope to control the commercial development of their inventions through patents, with the expectation of profit as well. But, like all creative people, they are more interested in the process of research and invention than innovation. Their investors want to see a big and rapid return on the cash they could have gambled elsewhere by selling the patents or the company holding them. This conflict between the art of invention and the commerce of investment is generally overlooked in the public conflict between the lone inventor and the larger corporations, but its resolution is every bit as stressful and as important for successful innovation.

While we often think of corporate inventors as less imaginative or interesting, they often act as lone inventors within the corporation. The corporate inventor accepts the benefits of a larger organization's financial, technical, and human resources. He or she has received a traditional technical education, and understands the need for a larger group of inventors and innovators to turn an invention into a commercial product. For inventions that are beyond the horizon or outside the agenda of the corporation's management, corporate inventors have to seek sponsors and build a team that will reduce the leader's ideas to practice. They often spend more time laying a paper trail justifying investment and often give up the opportunity to be first with a public demonstration. They also relinquish the possibility of entrepreneurial wealth in return for the security, opportunities, and professional recognition possible through corporate support. For most inventors, wealth is less important than the gratification that comes with having delivered a new and useful technology to society, and the opportunity to research and invent in a financially stable environment.

Despite the common devotion to invention, then, there are important differences between a competition for an invention and the Tour de France or a track race. Inventors do not start from a common line with equal forms of financial and technical support. They do not even start with a common definition of invention among themselves. Whether the context is small business or corporate, being first to a demonstration or even first to market is no indicator of ultimate success. Even assuming that they are, identifying a common milestone is not as simple as this makes it sound. The steps of television development, in utero, shall we say, overlap from paper to demonstration, making claims to parenthood more difficult.

Furthermore, the race is not over with the demonstration. In many ways it has just begun. Publicity can inspire or deter competition. No one knows if the patent will lead to a lab demo that works; if the lab demo will lead to a practical prototype; if the prototype can be scaled up to mass production; if a competitor will emerge with a more effective solution to the same problem. Credit for the claim of invention depends, then, on how the reader defines it. Is it an idea committed to paper, a patent application, or a reduction of that idea to practice? If invention is based on paper documentation, does anything matter besides being first to document the idea and file an application? Assume for the moment that patents rule. Do we consider extenuating circumstances that delay a patent application?

Patents are legal constructs, granted in the United States for technologies and techniques that are new, useful, and not obvious to others skilled in the art of that field of technology. Patent claims can be basic to a new technology, but they are most often rearrangements of existing technology or additions in the art to existing patents. The writers of the American Constitution did not create the patent system to be fair to all the people inventing the same solution to a problem. A patent assigns primacy to one inventor and provides him or her with the legal protection to encourage technical creativity and investment in innovation. With this protection the inventor is free to license or develop the invention and anticipate some monetary gains from the creative act involved. When two or more patents overlap in the claims for their inventions, a patent examiner determines their feasibility and the priority of one over the other. This is necessary more often than one might guess because of the frequency of instances of independent, simultaneous invention. Therefore a licensee enjoys the convenience of buying rights from only one inventor of the technology.

Most importantly, a patent is not the technology, although in court a contestant may have to demonstrate that the patent works as claimed. If one of the parties to a patent interference disagrees with the decision of the patent examiner, he or she can appeal, all the way to the Supreme Court, if desired. There, nine justices generally unfamiliar with science and technology may decide who is first. Assigning primacy for invention via the legal system means using a procedure intended to encourage the profit motive for innovation and possibly the decision of judges not "skilled in the art."

If an invention is the reduction to practice, how do we define practice and whose word do we accept that it occurred? Does that qualify the construction of a system that never extends beyond commercial novelty? This was the case with the first two systems to go to market in the United States and England. Does it mean reducing an electronic system to practice, and demonstrating it privately or publicly? Or does it mean turning another electronic system into what becomes ultimately the commercially successful system still used over sixty years later?

As we follow commercialization of these systems, other questions arise. First, how important is identifying a first demonstration if the inventors differ in their financing and sources of support? Corporate inventors often felt less urgency or were not permitted public demonstrations. How do we confirm that the demonstration took place, and should we privilege the others desperate for financing or free to publicize their work? Finally, is being the first inventor important if it leads to a technical and commercial dead end, or if it is not the basis for what consumers later embrace as a practical system? This and the following chapter should make it clear that, contrary to the popular attention paid to two of the four inventors described here, the reader can make solid cases for all four of them, and their many assistants, as television's inventors.

ELECTRONIC TELEVISION

After Campbell Swinton's 1911 proposal, a decade of further gestation followed before inventors began tackling the challenge of reducing any television system to practice. Few people read about the Englishman's electronic conception, and many researchers continued working unsuccessfully with Nipkow discs. Within those ten years, however, the technical demands of World War I, from 1914 to 1918, accelerated the development of vacuum tube electronics. European and American armies and navies underwrote applied research that made electronic and wireless communications more practical, widespread, and portable. Those improvements, the mass production of electronic components, and the education of thousands of men in the use of electronics fuelled the postwar boom in broadcast radio.

They also provided the tools for making some of the television proposals of earlier years a reality. These included durable amplifiers, better high-vacuum technology, improved cathode-ray tube (CRT) oscilloscopes, and more sensitive photoelectric materials. By the end of the 1910s, new inventors began applying for patents on video displays using CRTs. No one had solved the challenge of Campbell Swinton's CRT cameras, but the prospect of amplifying a video signal above electronic and thermal noise helped make it seem attainable. Yet, as Campbell Swinton commented after lecturing on "The Possibilities of Television" in 1920, it was "probably scarcely worth anyone's while to pursue it. I think you would have to spend some years in hard work, and then would the result be worth anything financially?" (Fisher and Fisher, 1996, 38–39).

Various inventors spent the 1920s on that challenge of reduction to practice, in which they applied their skills and experiences to the technical and financial challenges of making a new technology work. Several people proposed or tried to make electronic television systems, including Edouard Belin in France; Max Dieckmann, Rudolf Hell, and Denes von Mihaly in Germany; Kalman Tihanyi in Hungary; and Kenjiro Takayanagi in Japan. But none of them demonstrated a system before the two men discussed here.

VLADIMIR ZWORYKIN

The most promising electronic inventor was Russian-American Vladimir Zworykin. After four years of studying physics and working with Boris Rosing on television in Saint Petersburg, the twenty-five-year-old Russian ended his graduate studies in Europe in the summer of 1914 when Russia declared war on Germany and Austria-Hungary. In the conflict that resulted in the collapse of all three empires, Zworykin served his country in the application of radio technology on the Eastern Front. With the collapse of the tsarist monarchy, the socialist and communist revolutions that followed, and then a civil war, Zworykin proved himself a persuasive salesman, talking himself out of several life-threatening situations and into, out of, and back into the United States.

On returning to the United States in the spring of 1920, Zworykin joined other Russian émigrés at Westinghouse Electric and Manufacturing Company in East Pittsburgh. Westinghouse had steadily expanded its business from railroad brakes to electrical power systems to radio communications since the nineteenth century, and was a minority partner in the Radio Corporation of America (RCA), which administered the pool of radio patents for its parent companies. The advantage of working for such a large company is that an inventor can draw on its wide array of resources if the management agrees on the potential benefit to the company of the inventor's idea. The drawback is that the management may change or become disappointed in the progress of research and development, leaving the inventor isolated. Zworykin accepted this trade-off and spent his career navigating corporate politics on behalf of his projects. After some disappointments in the 1920s, he succeeded and enjoyed a productive career in which he eventually directed his own corporate laboratory.

Despite Zworykin's academic and government training, he showed a talent for commercial engineering when he devised equipment to mass-produce Westinghouse's most popular vacuum tube. He left a year later when asked to take a pay cut in the postwar recession or in a dispute over the patent rights to his work. He returned in March 1923, this time to Westinghouse's research department with his salary tripled on the recommendation of the company's patent department manager.

When asked to propose a research project, Zworykin suggested electronic television, for which he submitted a disclosure to his boss, Samuel Kintner, on April 4. He had been thinking about and discussing this with his Russian associates for several years, not having forgotten his work for Rosing over ten years earlier. By the spring of 1923, however, Campbell Swinton's ideas had been more widely publicized and Westinghouse and other companies were looking for ways to send facsimile images by wireless. Zworykin seemed to suggest that the next radio innovation could leapfrog facsimile and go straight to moving images. The notion of skipping an expensive step always appeals to investors in innovation. On October 8, Zworykin submitted his proposal for electronic television to the patent department, which, after two months of research and discussions with the inventor, submitted the patent to the U.S. government. Because in a dispute over a patent this basic Zworykin would have to show that it could work, he received the funding to begin ordering the necessary apparatus and parts. In the spring of 1924 he began working on the system full time.

The stimulus was perhaps less his lobbying than an article by Campbell Swinton published that April in the popular British magazine *Wireless World and Radio Review*. After reviewing his ideas and efforts to reduce them to practice, the Briton concluded that electronic television could only be accomplished by "one of the big research companies . . . who have a large staff and any amount of money . . . they would solve this thing in six months and make a reasonable job of it" (Abramson, 1987, 68). Zworykin began trying to fulfill Campbell Swinton's claim in June. The cathode-ray tube (CRT) for the display was not unusual, based as it was on an oscilloscope tube, and the camera was similar to Campbell Swinton's design. Zworykin's long cylindrical tube contained argon gas and a cathode at one end that emitted a beam of electrons. Electromagnets made this beam scan the backside of a three-layered plate at the other end of the tube. He and his technician made this plate of a thin sheet of aluminum foil, a coating of aluminum oxide for electrical insulation, and a layer of photosensitive potassium hydride.

On the other end of the tube, light from an image passed through a grid mounted close to the potassium hydride. The light triggered electrons continuously off its surface to the grid, conducted by the argon, which was positively charged, or ionized, by the electron beam. Each scan of the beam through the backside sent more electrons to the grid, where they formed an amplified instantaneous electronic image that drained off the grid to the transmitter.

Over eighteen months, Zworykin and his group built the specially adapted tubes, circuit boards, and other components for power, scanning, amplification, and display. Methodical trial and error, aided by current knowledge in chemistry and physics, led to new insights that in turn led to new patent applications and efforts to amend the 1923 application. The biggest problem continued to be the camera tube. The photoelectric surface proved to be as difficult to fabricate as Campbell Swinton indicated. The patent description for the surface that photons struck specified only a layer of photoelectric material. The lack of structure implicit in this description meant that the charges accumulated on any one spot diffused laterally, and the electronic image would quickly dissipate. Therefore the camera failed to store the charge built up between scans on the potassium hydride, making for an extremely inefficient transducer. By June 1924, Zworykin began using mica, a natural electrical insulator, between the aluminum foil and potassium hydrate. Over the summer he showed that "it was possible to receive on the phlourescent [sic] screen of the receiving tube a light line picked up by the transmitter [camera]" (Bannister, 2001, 83). A line on a CRT was not by itself unusual, since scientists and engineers had been using such tubes to show the trace of electronic signals for over twenty years.

To improve the photoelectric surface and apply Zworykin's skills to more immediately promising research, Kintner assigned more of his time to work on a more sensitive photocell with facsimile applications. Photocells became the subject of Zworykin's doctoral thesis in physics and gave him better insight on what would work in the camera tube. In a new application for a color television system patent filed in July 1925, Zworykin claimed that the layer of photoelectric material would be deposited as insulated "globules." This approach enabled preservation of a photoelectric charge in individual photocells but, because they operated only on the instant that the cathode beam scanned each of them, Zworykin's electronic camera stored no light from an image between scans.

Nonetheless, Zworykin had reduced Campbell Swinton's concept of an electronic video camera to practice. To sustain or increase his funding, Zworykin invited his next three superiors to a demonstration of his new technique in the fall of 1925. After spending all night repairing a blown circuit, he showed off his camera with its insulated mosaic of pixels and cathode-ray tube display. After a fruitless search for a mesh of Pyrex glass to replace the mica and insulate the potassium hydride elements, Zworykin captured the transmitted image in the mosaic, possibly in as a pattern of insulated rivets.

Westinghouse's senior staff was not impressed by the faint, fuzzy, distorted cross that appeared on the adapted oscilloscope. In addition, it was not an all-electronic television system, since Zworykin relied on motorized generators to enable the electromagnetic deflection of the scanning beam, the image did not move, and it is unlikely that the image was broadcast from the transmitter to the receiver. Nor did it help when the scientist spelled out the remaining challenges in the discussion after the demonstration. Senior Vice President Harry Davis quietly suggested to Kintner that he assign Zworykin "to work on something more useful" (Abramson, 1987, 81).

Certainly Zworykin's demonstration, which featured the world's first use of an electronic camera, was a long way from a practical, much less commercial, product. Home radio was still in its infancy. In 1925, the sound of broadcast radio included unregulated interference between stations, noise from the sparks generated by electric motors, and low-fidelity microphones and loudspeakers. Electronic television that produced a worse image than wireless facsimile was not going to expand production in the company's factories or add sales to the company's bottom line any time soon. Westinghouse transferred the television research and development program to their broadcast pioneer, Frank Conrad, who advocated an electromechanical approach. Zworykin spent the next three years working on facsimile systems. It was not television, but it kept him at the cutting edge of image transmission research and enabled him to finish his Ph.D. on photoelectric cells in 1926.

Meanwhile, the Patent Office declared that claims in Zworykin's 1923 patent application were in interference over two other patents. The patent had flaws that Zworykin later blamed in part on the challenge of explaining a new technology to lawyers in his second language. The Patent Office's examiner rejected it because of prior art; that is, earlier German and American patents and Campbell Swinton's articles invalidated many of Zworykin's claims for new and useful additions to television technology. The first conflict involved applications by not one, but ultimately six other inventors also making claims on the technologies and techniques involved with the image plate of an electronic camera. By 1929 the examiner of interferences had ruled in Westinghouse's favor in all proceedings on this issue.

The other conflict had to do with the electronic image created from the optical image of photons. Here is where the language barrier "impressed [Zworykin] tremendously with the importance of a good patent lawyer in the process of invention" (Abramson, 1995, 64). In Zworykin's patent the concept of an electronic image is necessary, for an electron beam cannot act on an optical image composed of photons. But the complexity of Zworykin's process, the quality of his English, and the novelty of the invention must have confused this issue for his Westinghouse lawyer; he failed to distinguish in the claims where the conversion from light to electricity took place. The challenges of claiming priority on these poorly written claims against the other patent applicants left Westinghouse and, later, RCA, defending and attempting to amend a weakly written patent.

Had the company been able to focus on Zworykin's cleaner 1925 patent and continue funding his work, perhaps there would have been less legal conflict later on. The concept of the electronic image would assume greater importance when one wanted to move it elsewhere in the tube. Zworykin and his staff at RCA would not begin working in that direction until the mid-1930s. The problem for them was economic, not technical, for legal credit for the concept and its manipulation belongs to a teenager named Philo Farnsworth.

PHILO FARNSWORTH

The youngest of our inventors was just fifteen when, in February 1922, he explained to his chemistry teacher at Rigby High School in Idaho his method of transmitting moving images electronically. A gifted child with an aptitude for physics, Farnsworth had been inspired at the beginning of the broadcast radio boom with the proposals for television published in Hugo Gernsback's and other popular science magazines. Virtually all of these drew on Paul Nipkow's spinning disks and the low-resolution imagery that might result if someone could make the technology work. Gernsback's *Electrical Experimenter* also contained an article explaining Swinton's electronic scanning approach along with the unresolved issue of suitable photoelectric materials.

Working by himself on an isolated farm, Farnsworth did the math and found it was impossible to scan for high definition with Nipkow discs. Looking at the alternating rows of cut hay in a field he had just cut helped inspire his solution of an electronic analog. Instead of scanning the image through Nipkow's disc in front of a light cell, he would create an electronic image from a photoelectric surface, and scan it back and forth, in a continuous pattern, before a single perforation containing an anode that led to the amplifier and transmitter. In the vacuum tube he described to his teacher, Justin Tolman, the light from an image was focused on a photo-electric plate. The photons from the light of the image stimulated instanta-neously an electron image, which would be pulled in its entirety by electro-magnetic coils through the vacuum to the aperture and anode. There, other electromagnets would pull the image back and forth, down and up before the aperture, which dissected it and successive frames so that the electrons in the images would hit the anode in series. That sequence would be channeled through amplifiers and a transmitter, to a CRT receiver, where the synchro-nized electron beam would reproduce the sequence on the phosphor face of the tube, line by line.

It was a brilliant solution by a teenaged genius, and Tolman kept his unique pupil's sketch of the system. Farnsworth's solution was not unique; Max Dieckmann and Rudolph Hell patented an image dissector in Germany in 1925. However, they failed to make a version that worked. The Idahoan could not afford to file a patent and had none of the resources necessary to make his idea a reality. Lacking advanced degrees or any useful connections, Tolman could only encourage Phil to "study like the devil and keep mum" (Godfrey, 2001, 14). This was excellent advice for someone trying to break into a major industry with a revolutionary invention, and Farnsworth fol-lowed it all the way to his first demonstration of electronic television in 1928.

Getting there required a uniquely American odyssey. It began with Farnsworth moving to Utah with his family, finishing high school, joining and dropping out of Brigham Young University and the U.S. Navy, and starting a radio repair business in Provo. It was not the ideal path for a career in, or gaining funding for, high technology. But after his father's death in January 1924, Phil was the oldest son and expected to support his mother and five younger siblings. Education took second place to feeding and housing his family for the next two years, and only work for some nonprofit fundraisers changed his fortunes. In the spring of 1926, George Everson and his partner Leslie Gorrell hired Farnsworth to help with their Community Chest campaign, and were impressed that he knew more about Everson's broken car than the local mechanics.

With the encouragement of Phil's future brother-in-law, they gradually teased out of the teenager his ideas for television. Like everyone else who heard him, Everson found Farnsworth's passion and knowledge riveting. The question was, what were the big radio companies doing? Surely they had teams of scientists and engineers working on the technology. Familiar with the technical literature in the university library, Farnsworth could assure them that no one was working on electronic television. Asked to estimate the cost of reducing his ideas to practice, he guessed $5,000. Like most inventors, Farnsworth underestimated the cost of turning ideas into reality. But Everson had $6,000 saved for a "longshot chance on something" and electronic television was "about as wild a gamble as I can imagine." If Farnsworth proved himself "it will be fine, but if we lose I won't squawk" (Everson, 1949, 45).

The three struck a deal and Farnsworth and his new bride Pemberley followed the investors to Los Angeles at the end of May. There the couple set up a lab in a Hollywood apartment where, with Everson and Gorrell following Farnsworth's instructions, they began building components for the system. Enough of these worked after three months and the expenditure of Everson's money for the investors to seek $25,000 to pay for another year of development. Farnsworth's conviction in the face of rising interest in Nipkow systems and the logic of his presentations won over executives from the Crocker First National Bank in San Francisco. For 60 percent of the shares in the organization, they would underwrite the $25,000 as well as Crocker Research Laboratory space that they already owned. Farnsworth, Everson, and Gorrell shared the other 40 percent.

The Farnsworths moved up the coast in September 1926, where Pem's brother Clifford Gardner joined them to learn glassblowing. They rented the second floor of 202 Green Street in San Francisco and found minimal equipment available. Farnsworth's team expanded to six people over the next twelve months, during which they mastered the techniques of glassblowing, the refinement and deposition of phosphors and photoelectric substances, magnetic coil winding, and tube assembly.

At the same time, on January 7, 1927, the twenty-year-old Farnsworth filed for his basic patent on an electronic system of television using the image dissector and the scanning of an electronic image. This was the beginning of a productive relationship with patent lawyer Donald Lippincott, who understood Farnsworth's ideas as basic principles in television circuitry and wrote them into very solid patents.

Despite the inventor's optimism in February 1927 that they could transmit a photograph right away, Farnsworth had to solve the problems of electromagnetic focusing of the electronic image on the aperture and the noise that developed with the amplification of the signal. Over the summer they built a four-element vacuum tube, a tetrode, by hand to increase the signal response. Ten days later Farnsworth methodically outlined the eight "Lines of Research" to be carried out "before offering any of our equipment for sale" (Godfrey, 2001, 32). On September 7, they transmitted a straight black line on a slide and then rotated it, the movement showing up against a green background of Willemite phosphors. "We've done it!" his wife recalled Phil saying. "There you have electronic television." This was not entirely true since they used motors to provide a high-voltage source and control the focusing and deflection coils. Nonetheless, Farnsworth could afford to wax more expressive in the telegram he sent to Gorrell: "The damn thing works!" (Fisher and Fisher, 1996, 151, 152). The dim, flickering, 1-inch display was not much different than what Zworykin had recorded two years earlier or what a standard oscilloscope could do, but it was an encouraging development.

They followed this up with photonegatives of Pem and Cliff on January 23, 1928, before accidentally discovering that exhaled cigarette smoke could be picked up on camera, although "care had to be taken to avoid a blistered nose" on the part of Gardner, the smoker, because of the heat from the arc lamp (Everson, 1949, 94). Over the winter, the bankers half jokingly asked Everson, acting as intermediary between Farnsworth and the investors, whether he had seen any dollar signs on the receiver. On February 13, 1928, Farnsworth transmitted, over wire, slides of a triangle and a dollar sign as silhouettes, illuminated from behind by an arc lamp. This was certainly cause for celebration, but the picture was not going to appeal to bankers looking for a commercial product in return for a second year's investment at $2,000 a month. Nor did it appeal to the General Electric Company's senior staff, one of who visited in March.

Farnsworth admitted that he could transmit silhouettes "with the amplifier system as it is, but I always get practically negative results when I attempt to show variations in two dimensions" (Fisher and Fisher, 1996, 154). Two months later, in early May, Farnsworth transmitted what he described as "real" pictures, with 2,500 elements over his system. After further inquiry by the San Francisco investors, however, GE concluded that "Mr. Farnsworth's scheme [did not amount] to very much" although he had "done some pretty scientific work with very limited facilities" (Abramson, 1995, 66). The "real" pictures did not impress his backers, either, who received a demo later that month and saw the dollar sign and smoke rings. By the end of May 1928, Farnsworth had spent over two years and $60,000 of the Crocker Bank's principals' money. They had already approached and been rejected by GE twice, and if the leading electronic company was not interested, it was time to stop throwing good money after bad and sell off Farnsworth's pending patents. It was not clear that motion picture footage would ever be possible, much less live images under the intense light of the arc lamp necessary to stimulate the dissector's photoelectric plate. The inventor disagreed. He believed, first, that he could show motion picture film on his system by the end of June and make their intellectual property even more valuable.

Second, Farnsworth had no intention of selling himself out to a larger company. His lab was going to survive and thrive on licensing income from patents, underwriting his research and the staff and equipment necessary to generate more inventions. In this approach he modeled himself on an idealized view of Thomas Edison's business model, neglecting the facts that Edison began as a contract researcher for Western Union Telegraph and that he had spun off his own companies to commercialize his inventions.

At the end of June, with no progress obvious, Farnsworth agreed with his investors to contribute to the lab's funding in proportion to his shareholdings. This meant he had to contribute half of the monthly costs by selling off some of his stock. This held off his increasingly jumpy investors until the end of the summer, when he scanned film clips with high contrast imagery. The twenty-two-year-old inventor showed "infinitely better" images from continuous loops of a boxing match and Mary Pickford's silent movie, *Taming of the Shrew*, thanks to a solenoid that improved the focusing of the electronic image on the opening of the aperture. Another project involved calculating what would constitute image resolution that was competitive with motion pictures. The group took photos of photographs through different densities of mesh screens and concluded that 400 lines would approximate the original image. To synchronize the scanning rate with the standard American 60-Hz power sources, Farnsworth also concluded that 30 frames per second (fps) would be more practical than the 24 fps used by motion pictures.

By September, Farnsworth had accomplished neither of these goals but demonstrated the system to the press and potential investors. Earle Ennis of the San Francisco *Chronicle* described it as "simple in the extreme" and a solution to "one of the major mechanical obstacles to the perfection of television" (Stashower, 2002, 144). The Christian *Science Monitor* called it "a radical step toward eliminating the problems inherent in the mechanical scanning disk" (Godfrey, 2001, 36). As far as they went, both were correct regarding the 8,000-element, 60-line, 20-fps video that they saw at the Green Street lab. On the other hand, Ennis described the demonstration of Gardner's smoking as "only 1.25 inches square with a queer looking little image in bluish light, that frequently smudges and blurs" (Abramson, 1995, 68). Flushed with the enthusiasm of youth and the reduction to practice, Farnsworth told reporters that commercial receivers "could easily be attached to an ordinary radio set and can be manufactured to retail at $100 or less" (Stashower, 2002, 144).

Whether Farnsworth's system was the path to an electronic system remained to be seen. For the tragedy of his triumph in 1928 is that the image dissector would never be used for live, commercially broadcast, television because of a flaw in the physics of the design. Like the Nipkow systems and Zworykin's first camera, it had no ability to store the electrons that created the image between scans. For example, if a frame of video contained 100,000 elements or pixels, the dissector could scan only one one-hundred thousandth of the light available to the frame through its aperture. The more pixels scanned in an image, the less light would be available. Thus the dissector needed enormous amounts of light to generate a picture, more than any human could practically bear while on camera in studio, and more than most climates can guarantee outdoors. Whatever Farnsworth did and said, he could not remedy what he explained to Ennis was "a matter of engineering" (Stashower, 2002, 144). The dissector would provide excellent grayscale pictures only in midday sunshine or through an arc lamp film projector.

Almost from the beginning of his efforts to build a dissector, Farnsworth recognized its insensitivity "was the bottleneck of television" (Everson, 1949, 112). It is less clear whether he ever acknowledged the lack of storage. He clung stubbornly to his first baby in television and spent the rest of his career trying to overcome its birth defect by improving the other components of the system. These often-brilliant efforts contributed greatly to the technology of signal amplification and added to his valuable patent portfolio, but never made the dissector suitable for broadcast television. The Idahoan's obsession and self-imposed isolation from other advanced researchers kept him from exploring other, more productive, approaches to scanning while others moved past him technologically.

NIPKOW TELEVISION

While Zworykin and Farnsworth worked in corporate and West Coast anonymity, two other inventors made Nipkow's approach work, 40 years after the German student conceived it. Charles Jenkins and John Logie Baird did so 3,000 miles apart and virtually simultaneously, which is less remarkable than it appears. Great minds often think alike when given the same ingredients with which to work.

We often think of invention as a young man's game, but the first American to demonstrate a working television system was 55 years old. A college-educated inventor from Ohio, Charles Jenkins applied his skills and imagination to new technologies just as the components to make them possible became available. In nearly 50 years of professional inventing, he accumulated over 400 U.S. patents. Jenkin's first interest was moving pictures in the 1880s, which resulted in his commercially successful Phantoscope movie projector a decade later. Then he took an interest in automotive technologies, patenting one of the first self-starter mechanisms and building Washington, DC's first tour bus in 1901. Jenkins made small fortunes based on his inventions and then lost them in the twists of market competition. He was philosophical about it: "The accumulation of great wealth does not seem to me an ambition which promises very great happiness" (Fisher and Fisher, 1996, 44).

Jenkins spent most of his time on moving pictures, and proposed transmitting them by wire in 1894 and 1913. He began thinking about transmitting them over the air in 1921, just as radio broadcasting broke out of its amateur market and Marcus Martin published *The Electrical Transmission of Photographs*. This book surveyed research in facsimile and television, including Campbell Swinton's electronic proposal. Jenkins rejected an electronic approach in search of a practical electromechanical technique based on an adaptation of the prismatic rings he used in his latest film projectors. That year he incorporated the Jenkins Laboratories to develop "radio movies to be broadcast for entertainment in the home" (Fisher and Fisher, 1996, 43). His first patent used two overlapping glass discs with prism-shaped rims of constantly varying cross section. The constant change enabled scanning of the elements of the entire image onto a light cell, which transmitted the video signal to a neon-gas lamp. This flickered in relation to the brightness of the signal, the light from which would then be projected through two more prismatic rings.

During 1922, with funding from the U. S. Navy and the Bureau of Standards, he used his rings to demonstrate radio-facsimile transmission of photos and images, leading the Navy to begin installing his equipment on ships for weather maps. The accomplishment, and perhaps the knowledge that Jenkins applied for eleven television patents that year, led a writer in *Scientific American* to imagine the transmission of sixteen images per second over improved equipment. At a rate equivalent to the frames projected in movie projectors, "we can turn a switch and see the latest play . . . or, by tuning out the play and tuning in the concert, watch the operatic singer as well as hear her" (Udelson, 1982, 26).

The system that Jenkins first tested in his laboratory in June 1923 represented a compromise with the Nipkow approach. Instead of his prismatic rings or the Nipkow spiral scan, Jenkins perforated 48 holes near the circumference of a disk and mounted a lens and differently angled prism in each. These prisms bent light from one of 48 strips, or lines, of the image and focused it on a light cell, sixteen times a second. At the receiver, a similar disc revolved synchronously and generated "a shadowy wave of the hand or movement of the fingers" (Burns, 1998, 202). Two years later, a private demonstration for Hugo Gernsback left the electronics promoter and publisher "under the influence of what I consider to be the most marvelous invention of the age" (Fisher and Fisher, 1996, 45).

The expense and difficulty in using and aligning the prisms drove Jenkins to the Nipkow approach. He did so reluctantly because the Nipkow system was not easily scalable. Two square inches of video display required a 36-inch disc; doubling the picture size would require a 6-foot diameter disc—"a rather impractical proposition in apparatus for home entertainment, even if it were possible to get power enough out of the house wiring to turn the disc up to speed" (Fisher and Fisher, 1996, 89). In addition, the Nipkow systems needed tremendous amounts of light to illuminate each pinhole as it scanned its line of image elements.

By the spring of 1925 Jenkins was explaining his system at the annual meeting of the Society of Motion Picture Engineers while failing to interest Westinghouse in his system. On a Saturday in June, Jenkins demonstrated to government officials his mix of lenses and Nipkow discs. General Electric provided the thallium-based photocell and neon glow lamp for the display, which was highly sensitive to changes in electrical current. His system broadcast on one channel from naval radio station NOF in Anacostia, Washington, DC, an image of a model Dutch windmill whose blades were moved by the breeze of an electric fan in motion. The group watched the video, which was "not clear-cut" but "easily distinguishable," on a 6" by 8" television receiver using a magnifying lens in Jenkins's laboratory at 1519 Connecticut Avenue Northwest. On a separate broadcast channel, an assistant at NOF predicted the motion of the blades' motion. *The New York Times* and *Washington Post* Sunday editions made this a page-one story, quoting the inventor to the effect that "the process would not be perfected until baseball games and prize fights could be sent long distances and reproduced on a screen by radio" (Udelson, 1982, 27).

The secretary of the navy looked for practical uses: "I suppose we'll be sitting at our desks during the next war and watching the battle in progress." Jenkins replied, "That's perfectly possible, Mr. Secretary." Was it television? Curiously, the inventor coined his own term, "radiovision," and his content "radiomovies." Television was aligned with the nineteenth-century technologies of telegraphs and telephones, which connected sender and receiver by cable. Having no illusions about the quality of the imagery, Jenkins highlighted the connection between his accomplishment and the modern entertainment media of the day. Unable to sell the system to either the government or the larger corporations, and protected by his patents, Jenkins began to build his own audience among radio enthusiasts seeking a new niche in which to pioneer now that broadcasting had become mainstream.

JOHN LOGIE BAIRD

Scotsman John Logie Baird matched Jenkins, albeit with far fewer resources beyond his remarkable mind. Baird was born in 1888 with an eccentric technical imagination, acquired perhaps when seriously ill as a two-year-old. After earning a technical certificate in 1912, he championed a variety of peculiar inventions and business ventures, none of which had any relation to any other and none of which proved profitable for long. He continued to be sickly and prone to cold feet, leading his London doctor to refer Baird to the healthier climate of Hastings in England's southeast during the winter of 1922–1923. There, "coughing, choking and spluttering, and so thin as to be transparent," Baird tried to invent a glass, rust-free razor and air-cushioned shoes with predictable results (Fisher and Fisher, 1996, 27).

With his savings steadily evaporating, in the spring the failed inventor-entrepreneur visited the town library and went for a long walk. He stopped at Fairlight Glen, overlooking the English Channel, where William the Conqueror led the Norman fleet that invaded England in 1066. There Baird's thoughts brought him back to vision at a distance, aided perhaps by an article on television in *Wireless World and Radio Review*. He had explored the concept of television as a teenager and now thought that, twenty years later, the necessary technological components existed to make a Nipkow-disc system possible.

The virtue of Baird's approach was that he had no idea how hard it would be to make television work, much less make it work well. He had, however, a tremendous passion that drove him in all his endeavors. It also attracted assistants willing to work or volunteer for someone with a vision. This attraction was a vital asset when the visionary had little mechanical aptitude or money. In June 1923, after a couple of months struggling with some local enthusiasts to turn a hatbox, some knitting needles, bicycle lamp lenses, batteries, and vacuum tubes into a Nipkow-disc television system, Baird advertised in *The Times* for a volunteer to help make it work. The major challenge involved the large selenium photocell, with its slow response time and a reaction to light that generated as much unwanted electronic noise as electronic signal. By July 26, he and his assistant had progressed far enough for Baird to invest in his first television patent. From July 1923, he began transmitting his first images over wire, so that after the first demonstration in November a reporter could typically anticipate the day when "we may shortly be able to sit at home in comfort and watch a thrilling run at an international football match, or the finish at the Derby" (Burns, 2000, 43).

Baird showed instead immobile silhouettes of crosses, letters, and other symbols through his spinning spiral of bicycle lamp lenses. The content had far to go, his camera and receiver were connected by cable, and his display offered a picture four inches square, perhaps 20 lines to a frame, and flickering at 20 frames a second. Nonetheless, when he began public demonstrations in January 1924, the national *Daily News* publicized his triumph in adding sight to sound only two years after the British Broadcasting Company (BBC) began formal radio broadcasting. Most usefully the recognition got him £50, the equivalent of ten weeks' skilled labor, from his father, and the approach of a prospective investor, publisher J.W.B. Odhams.

Before agreeing to Baird's offer of a 20-percent interest in his television in return for £100 and more business contacts, Odhams arranged for the donation of some expensive vacuum tubes from an engineer at the BBC and another demonstration to be evaluated by the engineer. With the improvements Baird could transmit moving silhouettes, but as might be expected, the engineer reported that Baird's technology was some distance from a commercial product, a view affirmed by the editor of the *Broadcaster*, a radio magazine. Odhams declined to invest but Baird gained his contacts anyway.

In February an article in *Radio Times* entitled "Seeing the World from an Armchair: When Television is an Accomplished Fact" suggested that the technology was the next big trend in wireless communications. After the obligatory references to the prospect of watching sporting events "or, for that matter, a battle," the author revealed that Baird and a C. F. Jenkins in the United States had both transmitted images through television systems (Burns, 1998, 152). In April a series of demonstrations for the press and the resultant articles led to the investment of £200 by Will Day, who supported Baird's efforts over the next twenty months in return for a share of Baird's first television patent.

Yet Baird himself did not even define what he had done as true television: that required the transmission of reflected light off a moving object, preferably over the air. Up to mid-1924 he focused on picking up images of silhouettes on a filmstrip so that the selenium cell had only to respond to the intense lamplight projected around the silhouette. The next step was to televise the reflected light off an object. As Baird calculated it, this required an increase of a thousand times more sensitivity by the camera to respond to far less light and show a passable gray scale. In June he wrote his patron that he had made the selenium cell "work by reflected light—that is, objects, not transparencies." His mistake was in adding confidently, "I feel quite certain that there are no insurmountable difficulties in the way of success" (Burns, 2000, 59–60).

Those surmountable difficulties took sixteen months to resolve. Day's investment disappeared and required additions as Baird filed for patents, bought batteries and tubes, and struggled to make his selenium cells show consistently a sharper, moving image from reflected light. He was, to a degree, the prisoner of his background and interest in mechanical and chemical engineering, for he resisted Day's encouragement to use photoelectric cells. The selenium was slow to respond but easy to use and relatively sensitive to light; the photoelectric cells responded quickly but were expensive and relatively insensitive. He had some electrical circuitry skills, for one of his 1924 patents applied differential calculus to replicate the light intensity from an image in the electrical current emitted from the selenium.

But Baird focused on the mechanism in two ways. First he added a serrated shutter revolving up to ten times faster than the Nipkow disc and a motorized ring of selenium cells. He thought that by pulsing the light on a series of cells he could nullify the lagging response. But this also required more synchronization and more voltage for the extra motors, and in July 1924 Baird nearly electrocuted himself in a 1,000-volt accident that made the local paper and resulted in his eviction from the apartment.

His second approach involved the cell itself and efforts to improve its sensitivity. This resulted in impractical and practical advances. After a dispute over Baird's progress or lack thereof in September, Day arranged for him to move his lab to a garret in Soho, London, closer to his patron, the following month. After settling in, Baird visited the Charing Cross Opthalmic Hospital in search of an eyeball. Having read that the eye's sensitivity to light was due to a fluid in the retina called visual purple, he convinced the chief surgeon to give him one. Baird's dissection was a messy failure. During 1924 Baird also began immersing electrodes in a colloidal selenium electrolyte. By suspending particles of selenium in a fluid, Baird increased the surface area of the photoconductor and its sensitivity by up to three orders of magnitude.

His focus on the light shutter, colloidal photocells, and the eyeball show how much the constant financial crisis affected the development of his system. The lack of money meant that the simplest solution was the best, regardless of its prospects for long-term development. When trying to demonstrate progress in a new technology to investors, however, that could be a virtue. Baird's persistence and the sheer fact of his televised images, unique in Great Britain, resulted in spasmodic publicity and haphazard reward. A neighbor in Hastings referred the owner of Selfridge's department store to Baird with the suggestion that television would be a "startling exhibit" for the store's anniversary (Burns, 2000, 75). In April 1925 George Selfridge paid Baird £60 for three weeks of transmissions of silhouette images from a system made of "[s]tring, cardboard, pieces of rough wood with Meccano parts, bits of bicycles and strange scraps of government surplus," as a friend of Baird's recalled (Burns, 2000, 73). Keeping the 3,000-volt system in order and dealing with a curious and affluent public exhausted the inventor, but it also resulted in the donation of several thousand dollars worth of batteries and vacuum tubes by two companies. But the inventor's demonstrations and sales pitches failed to attract other, larger investors. Reviews like that in the leading science journal, *Nature*, did not encourage it: "Mr. Baird has overcome many practical difficulties, but we are afraid that there are many more to be surmounted before ideal television is accomplished" (Burns, 1998, 158).

Thus Baird remembered his meeting with the general manager of the Marconi Wireless Telegraph Company as follows:

> "Good morning."
> "Good morning."
> "Are you interested in television."
> "Not in the very slightest degree, no interest whatsoever."
> "I am sorry to have wasted your time." [Baird had waited half an hour.]
> "Good morning." (Fisher and Fisher, 1996, 48)

Visits to newspaper offices turned out little better. Editors regarded the wild-maned Scotsman as another wireless "lunatic who should be watched carefully since he may have a razor hidden" (Fisher and Fisher, 1996, 48). Everyone who visited Baird's pathetic laboratories expressed astonishment and respect that he could show any transmitted video at all, but at the same time they could not see a commercial future for Nipkow disc television.

Frustrated by Baird's lack of communication and apparent lack of progress, Day wanted out. He had spent nearly £500 by May 1925 but expected Baird to raise additional funds as well as carry out the research and development necessary to a commercial product. This is an unreasonable expectation for most inventors, much less "a wretched nonentity working with soap boxes in a garret" as Baird described himself later (Fisher and Fisher, 1996, 52). But in June the two men incorporated Television Limited, so that Baird could sell shares in the company's stock, capitalized at £3,000, with the intention of buying back Day's investment.

Baird spent less time on stock pitches than with Stookie Bill, the head of a ventriloquist's doll that he used in tests. Any televised object needed a lot of light to show up on his neon lamp display, which replaced the light bulb behind the Nipkow disc in the receiver. The neon lamp was much more sensitive to changes in electronic frequencies sent by selenium cell, and on Friday, October 2, Baird finally obtained some gray scale on Stookie Bill's garishly painted face. It appeared as "a white oval, with dark patches for the eyes and mouth, [with] the mouth . . . clearly seen opening and closing" (Fisher and Fisher, 1996, 57). Baird raced down three flights of stairs to recruit a live subject: William Taynton, an office boy. Taynton retreated back downstairs when Baird went to the receiver, and the inventor had to pay him two and a half shillings to sit under the intense lights, thereby becoming the first person on TV.

Baird was excited with the success, nervous that a rival might beat him to a public demonstration, and down to £30 in savings. Some time in the next two months he reached back into his own entrepreneurial history and contacted Captain O. G. Hutchinson, a former rival in Baird's foray in the soap business. Hutchinson, an Irishman, was a natural salesman and liked the prospects of what Baird showed him. He bought Day's investment in December, and after the holidays proceeded to drum up Baird's technology.

On January 4, 1926, Hutchinson applied to the Postmaster General for a license to transmit television from four cities in the Great Britain using a 250-watt transmitter on a wavelength of 150–200 meters. Three days later, Baird showed Hutchinson's moving head to reporters for the *Daily Express*, and the *Evening Standard*. In between, Baird largely resolved the problem of the burning of the subject with the 16,000 candlepower necessary to resolve a 1-inch square image by reversing the placement of the light source and photocell. On Tuesday evening, January 26, Television Ltd. invited members of the Royal Institution and a reporter from *The Times* to a demonstration and forty people showed up. Dressed in formal attire, they clambered up the stairs to the landing outside Baird's attic rooms, waiting while groups of six entered to see the "faint and blurred" images of Stookie Bill and live talking heads. The "play of expression on the face" was apparent even to the reporter and led one television expert in the audience to admit, "Baird has got it. The rest is merely a matter of pounds-shillings-pence" (Burns, 2000, 90; Fisher and Fisher, 1996, 59).

Or so Hutchinson would have the public believe. He told one reporter that he had already ordered the construction of 500 "televisors," which would be sold for £30 and, when connected to a home radio, allow consumers to "look-in" as well as "listen-in." This was unlikely as he had yet to hear about the broadcast application and had no funds with which to pay a manufacturer. Perhaps the negotiations for a share of the business in return for the production of receivers explained the evaluation of Baird's system by an anonymous engineer in April 1926. Baird's Nipkow disc used 32 bull's-eye lenses from bicycle lamps, which scanned an area ten inches high and eight inches wide. This image was resolved down to a display one-ninth as

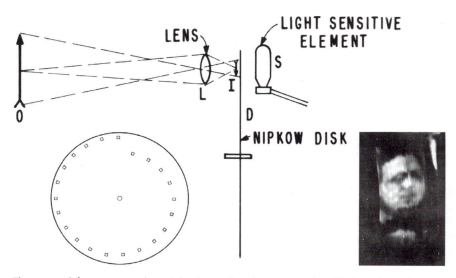

The second face captured on John Logie Baird's version of a Nipkow-disc television system was that of investor Oliver Hutchinson, in 30-line definition, in 1926. David Sarnoff Library, from Alfred Dinsdale, *Television* (1926)

large, although a magnifying lens could double it. Was the picture commercially appealing? The engineer could make out various facial movements but wrote that

> it would be very difficult to recognize an individual previously unknown . . . . The apparatus as now developed gives a crude image which is not even physically pleasant to view . . . . While the existing type of apparatus would undoubtedly achieve a temporary market the public would heartily tire of the results . . . . Those well-known personalities . . . would be scared off television by the present reproductions so that deserving developments later on would be hampered in securing support. (Burns, 2000, 92, 93)

More demonstrations and publicity followed, including a full page in *The New York Times* in March and a visit early in February from the Admiralty's leading television researcher. He failed to find a practical solution to the use of television in aerial reconnaissance by June 1926, but he was impressed by Baird's accomplishments and encouraged him to replace the selenium cell with a thallium sulfide photoelectric cell, which offered more sensitivity and a greater frequency response. In July, Alexander Russell, a fellow of the Royal Society, reported in *Nature* that his visiting group saw "living human faces, the proper gradation of light and shade, and all movements of the head, of the lips and mouth and of a cigarette and its smoke faithfully portrayed. . . . Naturally the results are far from perfect" (Burns, 1998, 165).

The next step was transmission over the air. Baird would have to transmit video over AM radio frequencies of 5 kHz; basing his calculations on reproducing the human face, he calculated that recognizable images could be picked up, broadcast, and displayed in a 30-line scan with 7:3 aspect ratio, sent twenty-five times every two seconds. He obtained an engineer's permission to use a BBC transmitter, and broadcast three times before a manager put an end to the unauthorized tests. With the approval of the Postmaster General for the experimental television broadcasts on July 15, however, Baird and Hutchinson could begin transmitting from the roof of Motograph House in Long Acre, London. To pay for it, Hutchinson filed to expand the company's capitalization to £9,050, or nearly $50,000. By the end of 1927 in a booming stock market, he had attracted 43 investors, and was preparing to make television a commercial reality.

Debate continues on the relevance of Jenkins's and Baird's work to the electronic systems that nations began adopting little more than ten years later. Their work spurred the first boom in television in the late 1920s and early 1930s, when the novelty of receiving moving pictures appealed to thousands of radio enthusiasts. Compared to the market of tens of millions of radio listeners and the quality of programming available to them, however, their embrace of the technically limited systems meant nothing. It had no bearing on the development of or marketing for electronic television some fifteen years later, save for awareness that the quality would have to be significantly higher in many respects. A useful analogy is the story of Lev Termen, who developed and demonstrated a 64-line system for his doctoral thesis at the Polytechnical Institute in Leningrad in June 1926. The Soviet Union abandoned mechanical television and purchased a complete electronic system from RCA in 1937. Nonetheless, those self-selected few working with these inventors saw live television before anyone else on earth.

3

Parenthood: Television's Innovation, 1928–1941

INTRODUCTION

John Logie Baird, Charles Jenkins, and Philo Farnsworth progressed as far as they had because of their own creativity and their relative independence from higher authority. The trade-off, of course, was their lack of funding and strategic planning. No one could argue with their claims to invention because they had demonstrated broadcast television first. In the long run, however, someone has to integrate a technological system with the rest of society. That means deciding what version of the system will become the standard that enables everyone to receive everyone else's signals. It helps to field-test the system to ensure that all components work together under operating conditions. Who will determine the technical standards that ensure that compatibility? How will manufacturers and marketers persuade customers that the time and money spent on this new system is worth giving up time and money invested on old ones? What will people decide to use it for?

Most inventors do not have to consider those questions except in the most hypothetical way. Even entrepreneurs, unless they are nationally ambitious, are more concerned with making a profit in the short term. The innovators who commit to making the technology a vital part of society, who pay for the research and development, the factories, the programming, the sales campaign, and service guarantees—they have to think through the most efficient way to make the system work socially and economically as well as technologically. This is innovation in its fullest sense.

Between 1928 and 1945, a wide variety of people contributed to television's innovation, converting it from a laboratory demonstration into a standardized, government-approved, commercial system with military applications. Beyond the engineers and scientists we usually associate with this process, investors, industrialists, technicians, factory workers, broadcasters, regulators, military officers, and advertisers all contributed. The number and diversity of people involved guaranteed that the evolution of the system would not be a linear process.

Versions of Paul Nipkow's electromechanical television dominated the publicity at first, thanks to the sophisticated efforts of inventors within larger companies, who joined pioneering lone inventors in advancing the cause of low-resolution video. The electromechanical approach came to naught because of its physical limitations; the onset of the Great Depression, which collapsed investment in the small businesses involved; and the commitment to fully electronic systems by larger companies. Attention turned to the systems under development by Philo Farnsworth, the Radio Corporation of America (RCA), and Electrical and Music Industries (EMI). Farnsworth tried to implement clever ways to circumvent the basic flaw in his camera, while expanding teams at EMI and RCA moved from strength to strength in refining and integrating a system ready for commercial broadcasting. Through government-controlled systems, the British and Germans began regular broadcasts in the mid-1930s, while rival manufacturers, broadcasters, and government regulators critiqued and delayed RCA's proposal. When the Federal Communications Commission finally approved a commercial standard in April 1941, military commitments cut short the introduction of home receivers. But the electronic inventions and innovations of the 1930s helped the United States armed forces during the war with improved communications, radar technology, and television-guided missiles.

NIPKOW SYSTEMS

As the lone inventors began attracting publicity for their accomplishments, and photoelectric and neon tubes became more available, others followed schedules serving corporate agendas or their own timetables. To be first with an invention means nothing except in one-sentence histories; to be second or third and supported by a corporation that can scale up funding, manufacturing, and marketing means having a technology that everyone uses. Was Nipkow the future of television? Why did it appear that everyone was embracing a technique innately limited by the amount of light available through the spinning discs?

If we look more closely at the motivations of participants, it becomes clear that only its passionate inventors, their investors, and a small number of radio enthusiasts took electromechanical television seriously. Larger institutions, from government agencies to major electronics corporations, insisted on higher standards of visual quality and content. Until inventors found a way to accomplish that, executives humored advocates in their organizations and pretended an interest in order to ride herd over the media attention accorded the novel technology.

GENERAL ELECTRIC COMPANY AND THE RADIO CORPORATION OF AMERICA

Although Jenkins and Baird demonstrated television earlier than anyone else, corporate inventors were not far behind. Because of their ability to flood the market once they committed to an innovation, they felt little need to compete with inventors starved for resources. At the General Electric Company (GE), Ernst Alexanderson added television to his research on radio facsimile transmitters in the spring of 1924. Best known for his work in high-frequency radio transmitters using rotary-disc alternators, Alexanderson soon became comfortable with the spinning wheels of Nipkow technology. Over the summer he urged GE's patent and engineering departments to pursue television as a natural extension of the company's interests in shortwave technologies. He told a vice president that "even moving pictures by radio, such as broadcasting of a boxing match is conceivable by the use of short waves" (Fisher and Fisher, 1996, 70). By December experiments showed that photoelectric cells responded to changes in light from a scanned image, and Alexanderson began to imagine displays 3 square feet in area. In mid-January 1925 his group gave a demonstration to company officials. They scanned a photographic transparency through a Nipkow disc and converted its grayscale levels to varying amounts of electric current. This was added to the signal from GE's radio station WGY in Schenectady, New York, which was in turn received and applied to an oscillograph mirror, the reflected light from which was divided by a rotating crystal, and projected on a small screen.

Having made this demonstration of principle, Alexanderson spent the next two years developing image transmission for facsimile and to a lesser degree for television, where the problem of inefficient transduction of light to electricity needed more research. In professional meetings, he spoke of shortwave radio making it possible to send and receive "letters and printed pages, and moving picture films, and ultimately to see by radio" (Brittain, 1992, 189). On the other hand, in December 1926 he noted that for acceptable television, transmission required 300,000 elements, or pixels, per second. This is the equivalent of a 10,000-pixel image sent thirty times per second. Alexanderson admitted Nipkow systems were not practically capable of scanning this volume of information, but "we are not willing to wait for a discovery that may never come" (Fisher and Fisher, 1996, 73). He knew of work on cathode-ray tube (CRT) displays, but these would not be appropriate for large-screen projection. Instead, his work showed that television could be introduced "with means that are in our possession at the present time" (Brittain, 1992, 190).

Why did Alexanderson embrace what even he knew was a limited technology? Perhaps it was the way the sudden rise of broadcast radio and electron-tube technology overwhelmed the years he spent making electromechanical point-to-point radio communications a practical reality. In any case, he followed his pronouncement with a demonstration in January 1927 in New York City of projected images televised from WGY up the Hudson River. Orrin Dunlap of *The New York Times* wrote that "it was a crude reproduction, but it moved" (Stashower, 2002, 122). None of the 500 engineers present, he opined, would disagree with Alexanderson's prediction that every household would have a television within a decade.

Nine months later Alexanderson told his superiors that his group had built a prototype home receiver. The converted radio console solved the challenges of synchronizing circuits by giving the viewer a tuning control. With practice, viewers could focus and recognize the individuals standing before the photocell camera in the laboratory. As usual, the image was small, not more than 2 inches square, but the orange glow of the neon tube allowed one to see the movement of lips and eyes as well as the inventor's standby, cigarette smoke. In January 1928, Alexanderson held a small press conference at GE's Schenectady labs to announce "the starting point of practical and popular television" (Stashower, 2002, 122).

WHAT DAVID SARNOFF THOUGHT

We might conclude from the accounts of Alexanderson's work that his superiors accepted his television initiative enthusiastically. One of the company's elite engineers who eventually received over 400 American patents, Alexanderson spent $500,000 on television-related research between 1926 and 1930. Although GE made most of its profits in lighting and power technologies, broadcast radio had been an unexpected spin-off of its research programs. GE channeled its business in wireless communications through the Radio Corporation of America (RCA), which its chairman, Owen Young, formed at the behest of the U.S. Navy in 1919. When RCA became the center of the new business of broadcast technology and its visionary vice president, David Sarnoff, pushed more applications in electronic communications, Young supported him whole-heartedly. Alexanderson served both companies and also enjoyed the interest of Young's protégé.

Sarnoff, however, muted his support. He was too far from Schenectady, part of a complicated hierarchy across two companies, and unhappy with the poor quality of the video display. Sarnoff continued to pay lip service to the efforts upstate and attended the January demonstration. There, in sweeping terms, he announced "that the radio art has bridged the gap between the laboratory and the home." Others had demonstrated television as a technical novelty, but now Sarnoff proposed "that the first step might be taken towards the development of television receivers for the home" (Stashower, 2002, 122).

What was the basis for his qualified claim? The group gathered around the two television receivers and watched—on the flickering, 48-line, 16-frame-per-second (fps), orange and black, 3-square-inch displays—an engineer announce that he was taking off his glasses, putting them back on, and going to blow smoke rings. Alexanderson's daughter remembered watching, on the family's receiver, a GE engineer "sitting around eating a chicken sandwich" on the tiny display (Kisseloff, 1995, 19). This was not the future of television and Sarnoff knew it. He continued to praise Alexanderson's efforts while waiting for someone to show how to make television a practical technology for the mass market.

Sarnoff also backed RCA's own explorations in Nipkow television. As the marketing and patent licensing agent for GE and Westinghouse radios in the 1920s, RCA enjoyed little support from its corporate parents for research in broadcasting. Its vice president started changing that policy during 1928. RCA's tiny Technical and Test Department began working on television late in 1927 and started receiving technical reports from Alexanderson's and Westinghouse television groups the same year. In 1928, the engineers at RCA's experimental TV station, W2XBS in Manhattan, bought papier mâché dolls of Felix the Cat. They mounted Felix on a turntable to test their Nipkow system for live motion and began building their own 48-line system, followed in 1929 by 60-line, 20-fps broadcasts. Thanks to publication of the schedule in amateur radio magazines and some newspapers, by 1930 up to 200 people with receivers as distant as Kansas tuned their tiny displays into the "dancing" cat, test patterns, photographs, and occasional live performers.

With the beginning of regular broadcasts from Schenectady on April 22, 1928, on 2XAF for video, and WGY for audio, television began making its way experimentally into American households. GE's press release announcing the Tuesday-Thursday-Friday schedule of afternoon broadcasts suggested that television enthusiasts might "pick up the signal and carry on independent investigations" (Fisher and Fisher, 1996, 94). Hoping to get one of the television receivers being tested, hundreds of people wrote in, asking to participate in Alexanderson's experiments. In August, maybe twenty people saw New York governor Alfred E. Smith accept the Democratic presidential nomination. On September 11, GE broadcast a 40-minute performance of J. Hartley Manners's 1899 melodrama, *The Queen's Messenger*. Izetta Jewell and Maurice Randall spoke into microphones placed in front of Nipkow cameras while a third camera watched male and female hands handle various props. WGY radio director Mortimer Stewart became the first television director, arranging for cuts between cameras while he monitored the broadcast on a receiver's tiny display. *The New York Times* gave it page-one coverage although the video images were "not always in the center of the receiving screen, and were sometimes hard on the eyes because of the way in which they flickered" (Ritchie, 1994, 21).

AMERICAN TELEPHONE AND TELEGRAPH COMPANY

At the Bell Telephone Laboratories of the American Telephone and Telegraph Company (AT&T), corporate support for television innovation was just as forthcoming. Newly formed in 1925 and located at 463 West Street in Murray Hill, Manhattan, Bell Labs represented the state of the art in electronic research. Backed by the financial resources of the company's telephone monopoly, it could afford to protect its interests and invest in any field that involved electrical and electronic communications if it suggested some telephonic application. The company had already filed for a substantial television system patent in 1917; in January 1925, the Labs' director, Harold Arnold, asked Herbert Ives to evaluate the prospects for transmitting video over telephone lines and to lay out a research program. Ives was a highly respected authority in optics and its applications in color photography, photoengraving, phosphors, and facsimile. He spelled out the challenges that so many had calculated before, and proposed "a very modest attack" on television with room for "expansion as new developments and inventions materialized" (Burns, 1998, 225).

Arnold funded him with $15,000 and in May Ives demonstrated his principles on a tabletop. Light from a carbon arc lamp shone through a photographic transparency and was scanned by a Nipkow disc, which stimulated a photoelectric cell at one end of the table. At the other, a neon lamp flickered synchronously with the changes in current from the cell, emitting changing amounts of orange light through the second, mechanically linked, spirally perforated disc. A Bell patent attorney reported that he could make out "with fair definition the features of a man's face . . . and that, when the picture at the transmitting end was moved forward or backward, or up or down, the picture at the receiving end followed these motions exactly" (Burns, 1998, 226).

This was not television, in Ives's opinion, and neither was the December 1925 demonstration of televised motion pictures or the March 1926 videophone conversation between two Bell Labs executives in the lab. Television meant covering "distances beyond any the eye could reach," which demanded the efforts of up to 100 engineers and took place late in April 1927. Secretary of Commerce Herbert Hoover, who oversaw the government regulation of radio, was televised live from Washington, DC, where he commented in a backhanded remark at Charles Jenkins's efforts, "The intricate processes of this invention could never have been developed under any conditions of isolated individual effort. I always find in these occasions a great stimulation to confidence in the future" (Stashower, 2002, 117).

His image and voice traveled by wire relay to two receivers in New York City. One was the standard neon glow lamp and Nipkow disc offering a 5-inch-square picture; the other was a unique neon tube folded back and forth to give a 120-square-inch display. The folds provided 50 lines to match the 50-hole Nipkow scanner, and a synchronized, rotating, electrical distributor that synchronously stimulated 2,500 connections to the 2,500 electrodes glued to the back of the tube. After Hoover's talk, the transmission switched to a television signal broadcast from Bell Labs' experimental station 3XN in Whippany, New Jersey. Onlookers watched another formal address by an engineer and vaudeville parodies of an Irish-American and an African-American. A year later, an observer from the British Post Office commented that the "very costly and elaborate piece of mechanism" gave "a clear and undistorted picture and the results obtained are undoubtedly far in advance of those claimed for by the Baird system" (Burns, 1998, 231).

Ives gained funding to keep "the Bell System on the map in connection with the onward course of television" through development of video conferencing technology and "whatever fundamental work would be necessary to safeguard the company's position" (Burns, 1998, 233). This open-ended endorsement by Franklin Jewett, president of the Labs, helps explains how Ives managed to spend over $308,000 between 1925 and 1930 on television research and development. Over the next three years he and his group explored and demonstrated improved photoelectric cells, outdoor pickups, projection and color television, and recording as well as further improvements in videophones. A demonstration of the latter in New York City in 1930–1931 attracted over 10,000 people. Erik Barnouw talked with his father on AT&T's first videophone system and remembered that his father's face on the screen looked "absolutely ghastly" (Kisseloff, 1995, 20).

NIPKOW ENTREPRENEURS

At the same time, Jenkins received the first experimental television license from Federal Radio Commission (FRC) in 1927 for station W3XK in Washington, DC Seeking to build a market for his system, he drew on the experiences of the early broadcast radio entrepreneurs. Through two of the first books devoted to television and frequent articles in Hugo Gernsback's magazines and QSL, the magazine for ham radio enthusiasts, Jenkins drummed up grassroots support for his technology, much as radio pioneers had done twenty years earlier. Jenkins started advertising television contests and selling by mail his $2.50 RadioVision kits. In July 1928 he started a regular schedule of "radiomovie" 48-line broadcasts of his animated silhouettes between 8 and 9 p.m. on Mondays, Wednesdays, and Fridays. Jenkins developed a mirror drum receiver and magnifying lens to expand the image to six square inches and promoted his "magic mirror" as able to tell "a pantomime picture story so realistic one's initial astonishment is lost in the fascination of the weirdly told tale" (Fisher and Fisher, 1996, 90).

As Jenkins expanded his video empire, actress Irma Kroman ran programming at his station in Jersey City, New Jersey, which opened next to his factory in January 1929. The silhouette films included "Little Girl Bouncing a Ball," "Little Black Sambo," and a passionate episode that involved, as one viewer wrote Jenkins, "much love at the breakfast table, many embraces, kissing and a goodbye and then the husband going back to the job swinging the pickaxe . . . and finally streaking for home and more embraces" (Udelson, 1982, 51). Kroman ran a lot of repeats, but she never received complaints because "there weren't that many people watching" (Kisseloff, 1995, 23). That year, Jenkins listed 30 loyal viewers between Massachusetts and Michigan.

Why were there so few? Despite the enthusiasts' hopeful analogies with the amateur broadcasting era ten to twenty years before, the differences were more significant. "Radio retailers are expecting a rich harvest to grow from the seeds of television now being planted," a *New York Times* reporter wrote in 1931 (Udelson, 1982, 56). But because television stations transmitted video signals over shortwaves and the audio signals over broadcast radio frequencies, TV entrepreneurs needed to reassure radio dealers and their customers that the new technology did not threaten to make radio obsolete. In addition, unlike the beginnings of broadcast radio, television was closely regulated from the start by the Federal Radio Commission, which had been founded to bring order to the amateur broadcasters of the early 1920s.

Nor was the content as varied as with early radio. The low sensitivity of the photocells made live transmissions almost impossible and films were harder to obtain than phonograph records. The movie industry was uncertain at the prospect of letting the public watch films in private instead of in a theatre. Representatives across the nascent television industry tried two forms of reassurance. First, they suggested that home television represented a different or unimportant market. Second, they argued that Hollywood would benefit by either the broadcast delivery of movies to theatres, as Alexanderson proposed and then demonstrated, or by the addition of live news events on the big screen to an evening's feature entertainment.

We do not know how far Jenkins thought Nipkow technology would progress through the support of the enthusiasts that he and other entrepreneurs cultivated. He had few illusions about the image quality, but his solution in 1929 was an even more impractical version of AT&T's display. Jenkins built an array, two feet square, of 2,304 tiny lamps, controlled by a motor-driven commutator to control their illumination in 48-line, 15-frame synchrony with the camera. He proposed a similar approach for the camera, using 2,304 tiny photocells. Like Alexanderson, he was an inventor trapped by experience and training, unable to switch from electromechanical to electronic technologies as others made them more practical.

Nipkow television became one of the last speculative crazes of the Roaring Twenties. Others jumped into the market, stimulated by publicity surrounding the GE and AT&T demonstrations. The maker of the neon tubes used in television displays, Raytheon Manufacturing Company, helped underwrite WLEX in Lexington, Massachusetts, which started broadcasting movies in April 1928. At Hollis Baird's transmitter in Boston, engineers and students from the Massachusetts Institute of Technology tried to pick up Red Sox baseball games across the street at Fenway Park on sunny days. In the studio, they also found that cigarette smoke transmitted well and had no objections to intense lighting. Ulises Sanabria demonstrated his Nipkow system at the Radio Manufacturers Association meeting in Chicago in June and then installed it at Chicago's WCFL. The station announced that it broadcast theatrical movies rather than the silhouettes some stations used. Not that it mattered much with 10-kHz bandwidth available for broadcast. At Sanabria's studio in Chicago, one job applicant put his hand before the light cell and recalled that on the receiver "you could just about distinguish your hand from a monkey wrench" (Kisseloff, 1995, 18).

In July *The New York Times* listed twelve television stations across the United States; nearly double that number, 21, had licenses by year's end. For all of these successes in sending crude monochrome facsimiles of reality, however, only fifteen stations were actually operating and GE, RCA, and Westinghouse operated seven of them. It was a far cry from the promotions of inventors seeking investors and publishers seeking readers.

In the absence of any serious commitment to Nipkow systems by RCA and its corporate parents, Jenkins took his system public in December 1928. Together with a group of investors, he incorporated the Jenkins Television Company for $10 million in stock to make cameras and receivers based on his patents, and license other companies to do the same. His new kits featured the components preassembled on an aluminum frame for $55, and in 1930 Jenkins offered a finished wooden cabinet television, similar to those made for radios, for $175, not including vacuum or neon tubes.

This represented the acme of Jenkins's efforts, for the Columbia Broadcasting System (CBS) bought his receivers for W2XAB's program testing in August 1930. The station showcased actors, comedians, chorus lines, celebrities, and the equivalent of vaudeville as staged by "Sally and Gene" three times a week until February 1933. In the depths of the Depression, CBS could not afford pointless experimentation, but announced that it "would resume our experimental transmission as soon as we are sufficiently satisfied that advanced equipment of a broader scope can be installed" (Ritchie, 1994, 30).

PROMOTING ELECTRONIC TELEVISION

Meanwhile, across Europe, in the United States, and in Japan, scientists and engineers proposed, patented, and began building electronic systems or its components. Max Dieckmann and Rudolph Hell had applied for a patent in Germany on an image dissector in April 1925, built several of the tubes, but failed to get an image out of one. Numerous inventors in France, Germany, and Japan adapted CRT oscilloscopes to serve as video displays.

The most successful was Kenjiro Takayanagi of the Hamamatsu Higher Technical Institute, who demonstrated a 40-line, 14-fps halftone video image of a Japanese character to the Japanese Society of Electricity in May 1928. Takayanagi had been refining CRTs since the early 1920s; like most other inventors in this field, he concentrated on electronic displays because of the technological maturity of the oscilloscope CRT and difficulties in making a camera tube.

Yet by 1930, no one but Vladimir Zworykin and Phil Farnsworth had demonstrated an electronic television camera. A technical and commercial dead end that became more obvious as the 1930s wore on, Farnsworth's image dissector nonetheless embodied a number of patents that anyone wishing to innovate an electronic system would have to license. The process by which that happened nearly destroyed the inventor, for he became enmeshed in business negotiations in three countries, all the while attempting to overcome the inherent flaw of no electron storage in his invention.

PHIL FARNSWORTH

Despite the publicity surrounding Farnsworth's demonstration on September 2, 1928, no radio company appeared to relieve the Crocker Bank investors of their share in the pending intellectual property. That fall, they stopped bankrolling the lab completely and a fire destroyed it at the end of October. Farnsworth and Everson convinced the investors to reopen the lab for more demos and the insurance settlement paid for its reconstruction. Over the winter the research group made incremental improvements. Farnsworth began developing his version of an electron multiplier tube. This device sent the few electrons entering from the anode aperture of the image dissector cascading down a series of metal plates. At each plate the impact released additional, secondary electrons, increasing the strength of the signal.

He was not the first or the only one to use this technique; Zworykin, for example, received seven patents for electron multiplication. But Farnsworth's multipactor, as he later called it, improved the image dissector's response, as did his application of the sawtooth waveform to the scanning process. Previously he and other inventors had used an electromechanical sine wave generator. Because the slope, or shape, of a sine wave changes constantly, so too does the velocity of the scan of a line of the image. All else being equal, this in turn varies the number of electrons emitted by the photocell and doubles the image because it scans the line in both directions. The sawtooth instead enables a constant velocity as it rises at one angle, stimulating constant electron absorption at the anode. Then it drops vertically, virtually eliminating the return scan and giving only a single image for each line. Construction of a vacuum-tube generator for the sawtooth signal by July 1929 made Farnsworth's system the first all-electronic television. Again, sawtooth scanning was not unique. After eleven years of interference proceedings in the Patent Office, Farnsworth won priority over rival applicants for its use in scanning an electronic image in television.

These improvements, as well as the new cesium oxide photocell, helped persuade Jesse McCargar, one of the Crocker Bank investors, to head a syndicate of investors in incorporating Television Laboratories as a privately held company on March 27, 1929. Farnsworth received 50 percent of the 10,000 shares held by the founders; the group expected to sell another 10,000 shares to other investors. Its purpose was "making discoveries and inventions of all kinds" and to "apply for, obtain, hold, own, use and acquire, sell, lease, and license . . . patents and patent rights in all countries in the world" (Godfrey, 2001, 37). The patents would also be used in manufacturing and operating a network of television stations, once the company received FRC approval and the necessary funding.

Farnsworth thought that the company needed about $250,000 in initial capitalization to continue research and expand, meaning that each new share sold would have to average $25 apiece. What would investors get for their money? Shareholders had no control over the company and saw no product or programming for sale. Instead, Farnsworth planned to license his patents to ten major radio companies and collect royalties.

There were several problems with this business model. He expected to license his patents as a package, when buyers would be interested only in cherry-picking those of basic importance and practical application. It was not yet clear that Farnsworth had rights to indispensable television techniques and therefore the leverage to force package licensing. In addition, his demonstrations, while improving over time, had yet to show a significant advance over Nipkow systems. In any case, until the patents were granted, the first three of which he received by August 1930, no one would buy anything from Farnsworth.

This raised serious cash flow problems when the stock market crashed in October 1929. With it popped the speculative bubble of investment in television and interest in the Television Laboratories' promotions. Farnsworth spent over $19,000 on development and patent costs in 1930 with no sign of a return in the foreseeable future. By the fall of 1930 McCargar was in a financial squeeze. He had resigned from Crocker Bank to become an independent financial consultant and sell shares in the new enterprise; neither of these occupations paid many bills after Wall Street's collapse.

Desperate to cut costs, McCargar fired the staff, which for the second time agreed to work for free. As the Great Depression set in during 1931, Everson sold only 1,012 shares at an average of $27. Thus hopes rested on attracting a corporate investor, and Farnsworth and his backers met with or gave demonstrations to eight radio-related companies, including Westinghouse and RCA.

SARNOFF, RCA, AND THE FUTURE OF TELEVISION, 1928–1930

Between 1928 and 1930, a great deal changed at what one critic called the "Radio Octopus." In particular, thirty-nine-year-old David Sarnoff became president of RCA on New Year's Day, 1930 and began implementing plans to extend the company's grasp over the new domain of television. But who was this man, and what was this company?

Exceptionally bright, and toughened by immigration from Russia to New York City's Lower East Side and the need to support his family from childhood, Sarnoff started working for RCA's predecessor, the Marconi Wireless Telegraph Company of America, in 1906. By the time General Electric Company bought the company and turned it into RCA in 1919, he learned everything he could from the engineering and marketing staffs as he rose from office boy to commercial manager, driven by an ambition to improve wireless communications in his adopted country. With the innovation of Lee De Forest's three-element vacuum tube, Sarnoff connected the ability to broadcast a continuous, electromagnetic, audio signal to mass audiences eager to hear the best music and the latest news. The government-approved pooling of radio patents in RCA and the grassroots boom in broadcasting after World War I gave him the tool and the market to realize his vision.

Yet RCA barely benefited from its monopoly of radio patents. GE and Westinghouse produced radios and RCA marketed and sold them. Coordinating engineering and production between the corporations left RCA behind in offering new features on which it held patents. The few barriers to making radio receivers resulted in sales booms and busts. Until RCA began the National Broadcasting Company (NBC) and started enforcing the licensing of its patents in 1926–1927, the erratic quality of the receivers—and the programs broadcast through them—increased consumers' distrust of the new technology. Consequently it is arguable whether the company made a net profit in radio sales while the industry gross approached a billion dollars a year by 1930.

Sarnoff spent the late 1920s trying to gain the RCA's independence from GE and Westinghouse. For it to produce the innovations Sarnoff envisioned as socially useful and commercially profitable, he argued successfully that RCA needed its own research and manufacturing resources. Continued negotiations led to the purchase of Victor Talking Machine Company in 1929, which gave RCA its first factory.

Now, with the resources, responsibility, and leadership to innovate the next generation of electronic broadcast communications, Sarnoff was determined not to repeat the mistakes of radio entrepreneurs in the 1920s. Unlike them, he understood that the successful innovation of television depended on more than the "black box" of the technology. RCA possessed a monopoly of radio patents; ran a research organization of scientists and engineers; manufactured radios, vacuum tubes, and related products; and owned NBC, which operated the largest network of radio stations across the country. This operation would enable the research and development of a commercially acceptable television system; the production of dependable transmission equipment and consumer electronics; the production of consistent and broadly appealing networked programming; and the market power to drive reluctant radio and broadcast industrialists into an innovation that threatened their status quo.

Or so Sarnoff anticipated. He had begun predicting the innovation of television in 1923 and, in a speech at the University of Missouri a year later, asked his audience to "think of every farmhouse equipped with . . . a screen that would mirror the sights of life. Think of your family, sitting down on an evening in the comfort of your own home . . . not only listening to a sermon but watching every play of emotion on the preacher's face as he exhorts the congregation to the path of religion" (Lyons, 1966, 207–208).

But he had not been impressed with the Nipkow system as the way forward. In November 1928, Sarnoff wrote in *The New York Times* that RCA was not opposed to television, only to the low-resolution standard proposed by the Radio Manufacturers Association a month before. He predicted the broadcasting of movies and live action in three to five years. In January 1929, Vladimir Zworykin explained electronic television to Sarnoff and offered to build at least a laboratory system in two years for a price he estimated as $100,000. Sarnoff approved the investment through Westinghouse until he could arrange the transfer of GE and Westinghouse's electronics staffs to RCA.

With the unification of electronics research and development at RCA late in 1929, technical staff from Westinghouse in East Pittsburgh, GE in Schenectady, and RCA in New York began packing their families and slide rules somewhat reluctantly for the move south to Camden. There, at the new RCA Victor Company, they unpacked the tensions between rival groups, technological approaches, and corporate cultures. Even without the threat of the Depression, this was a tense period for television at RCA. Some 45 engineers, technicians, and scientists converged to work in television by the transfer date of April 1, 1930. Zworykin had an informal deadline to show television's electronic future in less than a year; Alexanderson pushed his Nipkow system for immediate commercialization; and RCA's team was inclined to support Alexanderson based on its experiences. Looking in from New York was Sarnoff, whose reformed company came under assault by the nation's economic collapse and by the U.S. Department of Justice for violating antitrust statutes.

In May 1930, Sarnoff was handed a summons with the formal charges against RCA and its parent companies and patent pool members. He spent the next 29 months negotiating to keep RCA intact while also fighting to keep it solvent. Stuck with an outdated factory in Camden tailored to the production of talking machines, Sarnoff and RCA focused on the development of RKO Radio Pictures, increased advertising on NBC, and the prospects of television as a new product for whose unique features consumers would pay a premium in a depression.

THE KINESCOPE TELEVISION DISPLAY, 1928–1929

The first order of business for Zworykin was the display, for technological and practical reasons. It was going to take more time to build the first electronic camera that stored information and provided images from reflected light; that had never been done before. It also made sense to show his patron how the brightness, resolution, and size of the CRT provided a larger, sharper video image than a neon lamp, Nipkow disc, and magnifying lens ever could. By the end of June, Zworykin wrote in his quarterly report of a demonstration held on May 10 and of his expectation that the CRT would "be adapted for mass production" (Bannister, 2001, 88).

Besides, engineers and scientists had been improving the oscilloscope for over thirty years, and several had made significant advances in adapting it for video, including Ray Kell at GE. In November 1928, Zworykin had toured RCA licensees and laboratories in England, Germany, and France to report on the latest developments in facsimile and television. In Paris, through the recommendation of a colleague from before World War I, he visited the Laboratoire des Etablissements Edouard Belin.

Belin had worked on television since 1903. His latest development involved the electron beam of a CRT. There were three approaches to focusing an electron beam: using gas in the tube to help conduct the electron beam, as in the most popular oscilloscopes; using electromagnets outside the tube and a vacuum inside; or using electrostatic deflection inside the tube. The gas-filled tubes tended to wear out the electron-emitting material coating the cathode and varied in brightness as the beam scanned larger display screens. Magnetic focusing, as Farnsworth applied it at the time, gave a dim image because the electron beam could not be accelerated. Electrostatic focusing had its own problems, but at Belin's lab Pierre Chevallier and Fernand Holweck solved many of them by completely evacuating the gas inside the CRT. They also drew on and developed further the principles of electron optics. First posed in Germany two years earlier, electron optics offered mathematical analogies between the behavior of light and the behavior of electrons between electrodes. The Frenchmen used this technique to model electron behavior in a vacuum tube and adjust the voltages between the deflection plates and the anodes in specific ratios. This greatly improved the focus of the beam, and therefore the definition of the image, which promised the practical use of a CRT display.

Zworykin bought one of the Belin tubes and brought it back to the United States on Christmas Eve, 1928. Once Sarnoff gave him permission to move forward, he began improving the French CRT by rearranging the deflection plates between the first, focusing, anode and the second, accelerating, anode, which consisted of the metalized coating of the CRT's bulb around the green fluorescent screen. The second anode accelerated and focused the modulated electron beam and also drained electrons off the screen once they had stimulated the phosphors. By April 1929 Zworykin had a relatively inexpensive, bright, higher-definition CRT that RCA would trademark as the Kinescope, where one could see an image clearly under daylight conditions. By August he could receive a television broadcast on it. Westinghouse built six television receivers using the 6-inch Kinescope that fall; on the one in his house Zworykin watched 60-line, 12-frame film loops transmitted by KDKA between two and three o'clock in the morning.

Zworykin's ideas were not unique. For the Kinescope's electrostatic operation he drew not only on the Chevallier-Holweck work but that of Roscoe George and Howard Heim at Purdue University. Eight years passed before Zworykin's patent application issued, and RCA bought other inventors' patents on CRT televisions if its technical and legal staffs agreed on their value.

These maneuvers reflect the standard practice of any monopoly. If the company's staff does not receive the basic patents in its products, then it must buy or exclusively license those patents from others to maintain its reason for existence. Although many monopolists use their grip on intellectual property to limit development, Sarnoff saw the patent pool as a tool for controlled revolutionary innovation. Consumers would receive steady improvements in their current products while RCA's engineers worked to ensure that revolutionary systems functioned as promised, out of the box. In addition licensees enjoyed the resources of the company's technical staff from which they received engineering updates and consulting support while making their own improvements. Sarnoff believed that the process ensured predictable and renewable economic growth without the boom and bust cycles that plagued the radio industry in the 1920s.

FARNSWORTH AND RCA, 1930

When RCA received an invitation to visit Farnsworth's San Francisco lab in April 1930, it could not have come at a more inconvenient time for Zworykin. He had just arrived at RCA's shabby Camden factory, was still under contract with Westinghouse for television projects at least until July, and was at the same time competing with Alexanderson for television's future at the company. The visit was little different in procedure from Zworykin's tours in Europe, which he made throughout his career. He evaluated the demonstrated technologies and reported on their value to RCA. If necessary, RCA bought the patents or licensed them exclusively as part of its own patent pool. Further, if a researcher at the lab appeared to have skills useful to RCA and an interest in relocating, RCA hired him.

Zworykin arrived at Green Street in San Francisco in mid-April. He spent three days with Farnsworth and his group, admiring the quality of craftsmanship in the image dissector tube, and Cliff Gardner's technique for fusing a flat Pyrex-glass disc to a glass cylinder for the dissector. No glassblower at Westinghouse or RCA thought it was possible. "This is a beautiful instrument. I wish I might have invented it," he commented (Everson, 1949, 126). The lengthy visit, the compliments, and dinner at Farnsworth's house all suggest that Zworykin got along well with the younger inventor. His report praised the dissector and sawtooth scanning technique for, whatever the quality of images—and Zworykin received a screen shot of Pem Farnsworth as a reminder—they were better than anything he had yet accomplished.

In June RCA sent out a patent lawyer and its director of research and development, Albert Murray, to review Farnsworth's patents and see a demonstration. Echoing Alexanderson at GE, they concluded that RCA could work without Farnsworth. After reading Zworykin's report, Murray opined that "Farnsworth has done some very clever work, but I do not think that television is going to develop along these lines . . . . Farnsworth can do greater service as a competitor to the Radio Corporation group" (Abramson, 1995, 91).

MAKING AN ICONOSCOPE CAMERA WORK, 1924–1931

He already had in two ways. First, Zworykin sent an 800-word telegram to Westinghouse requesting the construction of several image dissectors. He stopped in East Pittsburgh and picked up the new dissectors en route to Camden where his group was still settling in. The Westinghouse dissectors were up to a hundred times more sensitive than Farnsworth's and Gardner's, thanks to the photoelectric material patented by GE's Lewis Koller. For the next eighteen months, Zworykin's team experimented with dissectors, using them as their primary source for electronic imagery while they struggled to make a camera tube using the storage principle.

Second, two weeks later, on May 1, Zworykin applied for a patent on a camera tube that explained explicitly the storage principle to which his earlier patents alluded and how to include it in his camera tube. The visit to San Francisco revealed a working camera, but one that Zworykin recognized as fatally flawed by its inability to store the light between scans. Despite the references to storage elements in numerous claims of his 1923 patent application and the reference to insulated globules on his 1925 patent, Zworykin and his lawyer had never explained the advantages or means of storing charge in proportion to the light signal. Lawyers for Westinghouse repeatedly tried to insert Zworykin's later claims in the 1923 application, up through his 1931 application regarding a mosaic of pixelated targets with storage electrodes. It ended up in interference with two other applicants; when RCA won one and lost the other in 1936, it bought the winning patent.

Based on Campbell Swinton's proposal, Zworykin's Westinghouse cameras lacked the ability to store light as electrons between scans. As with the kinescope, Zworykin looked to others' work for guidance to a patentable solution. Before 1930, five other inventors referred to the benefits of charge storage in patent applications and offered solutions. Harold McCreary of Chicago was the first to suggest connecting electrodes to a mosaic of sensors, in his patent for electronic color television in 1924. The imaginative Hungarian Kalman Tihanyi also began proposing methods of storing photoelectric charges in a camera in 1926. No European patent office approved his applications, and given his itinerant career with military patrons in several countries, we might suspect that his array of electronic solutions for many problems proved impractical. Within those solutions, however, lay several useful concepts that Zworykin adapted.

Late in 1929 Gregory Ogloblinsky, chief engineer of the Belin lab, joined Zworkin's ten-man team and began building an image plate using insulated rivets as proposed by McCreary, but coated with a cesium-silver oxide photoelectric blend invented at GE. This mosaic was the first effective use of charge storage, where the scanning beam discharged the charge capacitance on each pixel relative to the amount of light shining on it between scans. Ideally this would maximize the transduction, or conversion of light to electrons, but amplification of the video signal by vacuum tubes added immense amounts of noise. Nonetheless, the resulting 12-line image "proved to be quite promising," as Zworykin wrote in an internal report. "A rough picture was actually transmitted across the room using cathode ray tubes for both transmitter and receiver" (Abramson, 1995, 82).

THE END OF ELECTROMECHANICAL TELEVISION, 1930

Meanwhile, Alexanderson forged ahead with his mirror-drum and Nipkow scanners. Although it began to look impractical for the home, the inventor also applied a version to theatre television. As Sarnoff had recently organized RCA's latest venture, RKO Radio Pictures, to exploit GE's soundtrack technology, this was a sensible strategy. Alexanderson first demonstrated it in a transmission from the Schenectady labs to the Proctor Theater downtown a mile away. In May 1930, the small audience saw six square feet of dim, streaky, muddy video of vaudeville acts and an orchestra projected from a 48-line Kerr cell. Three weeks later RCA announced that it would install television projectors in RKO theatres nationwide. But the Depression and the ongoing costs of converting to talking pictures, as well as the crudity of Alexanderson's system, ended ambitions for theatre television for another ten years.

Zworykin's group could not fabricate one of his new cameras in time for the shoot-out with Alexanderson's system on July 15, but it did not matter. The contest, held in Collingswood, outside Camden, showed that the Kinescope offered prospects for a bigger, brighter, and higher resolution picture than Nipkow systems could ever provide. From this point forward, ninety percent of RCA's funding for television research went to the electronic system. Two days later, Zworykin applied for his first RCA patent, and members of Alexanderson's team began transferring to Zworykin's expanding group. By September, all RCA television research was organized under Zworykin.

In RCA's annual report for 1930, Sarnoff summarized the past and the future of television as he saw it. Television that must "broadcast regularly visible objects in the studio, or scenes occurring at other places through remote control" to home receivers "that will make these objects and scenes clearly discernible in millions of homes," all without using "rotary scanning discs, delicate hand controls and other movable parts" (Stashower, 2002, 168). While it represented a final rebuke to Alexanderson and other Nipkow optimists, Sarnoff also set a high bar for Zworykin's system, one that required another eleven years of technical development and regulatory debate to become a commercial standard.

Meanwhile the Federal Radio Commission held hearings on the appropriate bandwidth for television. The wider the band of frequencies allowed for each broadcast channel, the more video information can be transmitted. The higher the frequency of a transmission, the more directional it becomes and the more power it requires to broadcast over a given distance. Radio programs were transmitted over 5-kHz channels using frequencies of hundreds of kilohertz. In the early 1930s, there were few tubes or antennas capable of transmitting, amplifying, or receiving frequencies hundreds of times higher. Thus when Farnsworth appeared in Washington, DC, in December 1930, and rejected the intuitive assumption that higher definition television required more frequency bandwidth for transmission, many engineers rejected his claims. He asserted that a new tube that he had developed made it possible to send a 300-line video image using 2.5 million pixels per second on a 6-kHz channel.

AT&T's Herbert Ives responded that if "Mr. Farnsworth is doing what he says he is doing, we simply do not know how he does it" (Fisher and Fisher, 1996, 212). Video signal compression blossomed only with digital television's innovation in the 1990s, and while Farnsworth retracted his claims, he and his backers pumped up the publicity for their system throughout the spring. Television Laboratories had a couple of suitors by then, one of which commissioned an independent review. This time, again using film and high-contrast photographs, Farnsworth showed 200-line pictures whose "improved picture details . . . are very noticeable" compared to the Bell Labs and RCA broadcasts at 60 and 72 lines per frame (Godfrey, 2001, 51).

This praise attracted Philco, a rising radio manufacturer. With Sarnoff predicting the arrival of television in the next two years, Philco's owners resented depending on RCA for access to the next generation of consumer electronics. Sarnoff's confident assertions helped drive them and other corporate suitors to San Francisco in the spring of 1931, enabling Farnsworth to negotiate an agreement that gave him what he wanted. With Philco, Television Laboratories retained its intellectual property and received funding for research and development billed against future royalties. Farnsworth had to move his family and most of his team to Philco's factory in north Philadelphia, but he also moved closer to the technical, media, and financial resources of the northeastern United States.

In these circumstances Sarnoff himself visited Farnsworth's laboratory. *The New York Times* promoted Farnsworth's work during his Washington testimony and in May 1931 *Radio News* published the first extensive explanation of Farnsworth's 200- to 400-line system. This included the blurry off-screen photo of Pem and rumors of discussions that suggested commercialization by the year's end. Since Farnsworth received his first patents after three years with minimal delay, while Zworykin's first, 1923, application still stewed through a series of interferences in the Patent Office, it made sense for Sarnoff to assess a likely rival for the intellectual property on which his and RCA's ambitions depended. That month, Sarnoff visited RCA's RKO studios in Hollywood and then traveled north to Television Laboratories.

As fate would have it, Farnsworth was on the east coast, finalizing his deal with Philco, when Sarnoff arrived on May 15. George Everson gave him the tour in which RCA's president showed less interest in Farnsworth's displays than his camera. Sarnoff offered $100,000 for the company's patents and Farnsworth's services. The inventor rejected the offer, which neither covered the costs incurred over four years nor offered the individual freedom he was obtaining in the deal with Philco. This opening bid led to no further negotiations and Sarnoff departed, asserting that RCA could do without Farnsworth or his patents. A month later the inventor, his family, and most of his staff packed for the train trip east.

For the next two years, Farnsworth's team struggled to adapt to the demands of a large manufacturer while they and five ex-RCA engineers hired by Philco confirmed that the image dissector was not a practical device for live television. They also got their first look at RCA's experiments with storage tubes, picking up the experimental broadcasts from across the Delaware River on the new Philco receivers.

BIRTHING THE ICONOSCOPE, 1931–1933

The news was not good. By May 1931 RCA's team had become frustrated by the work on a two-sided camera tube, where light from an image shone on one side of a photoelectric mosaic while an electron beam scanned the other. Now Zworykin and his team tried another of Tihanyi's proposals and suggested a single-sided approach, where the light from an image and the scanning electron beam both struck the same surface. It was a counter-intuitive suggestion, since one could expect the electron beam to destroy the photosensitive surface's storage capacity.

To his team's surprise and for reasons never satisfactorily explained, a tube with a surface consisting of insulated silver rivets made by Harley Iams, Les Flory, and Sanford Essig gave promising results. The next challenge was a practical surface, since scaling up the number of rivets promised to be expensive. The group tried various ways of depositing and insulating silver particles as pixels, but in July Essig made a serendipitous mistake by over-baking the image mosaic to accomplish the breakthrough. By November RCA filed the first patent for Zworykin's Iconoscope, and over the next two years his seven-man group refined the techniques for fabricating the components and assembling them into a practical tube. RCA announced Zworykin's breakthrough in June 1933, leading to publication of an article in multiple technical journals and a European tour for the scientist.

THE PROBLEMS WITH PATENTS, 1932–1938

RCA also began buying up the patents of international inventors of numerous, unbuilt, single-sided, camera tubes. It could not, however, buy up Farnsworth's image dissector patent position. Instead it tried to disallow the claim to an electrical (i.e., electronic) image in the inventor's 1930 patent, number 1,773,980. In May 1932, the Patent Office declared an interference claim on behalf of Zworykin's still-pending 1923 application against Farnsworth's patent for a television system using an electronic image. RCA's lawyers asserted that Zworykin had explained his concepts to fellow Russian émigrés as early as 1919; Farnsworth's lawyers claimed that Zworykin "had shown no conception of an operative device prior to Farnsworth's patent" and therefore his application could "not constitute a reduction to practice" (Godfrey, 2001, 74). The case against the legitimacy of Farnsworth's claim rested on three issues: the date of conception, for which Farnsworth's lawyers produced both his teacher Justin Tolman and Farnsworth's 1922 drawings; the date of reduction to practice, meaning here the application date for the relevant patent; and the date of the device's operation.

On the first test, RCA's lawyers convinced the patent examiner that Tolman's testimony was "vague and incomplete," the 1922 sketches too

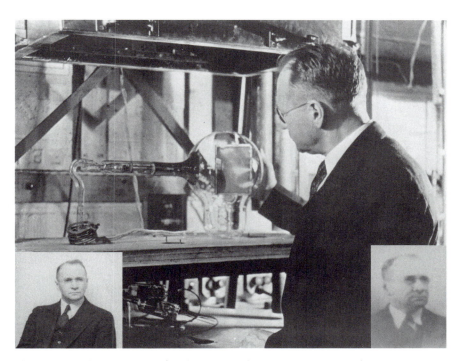

Vladimir Zworykin examines a freshly evacuated iconoscope camera tube in RCA Victor's lab in Camden, New Jersey, sometime in 1934–1935. The first video imaging device to store light as electronic charges between scans, the iconoscope improved television-camera resolution by an order of magnitude. Compare the photo of Zworykin on lower left with the televised image on the lower right, with John Logie Baird's Nipkow-disc image on page 36. David Sarnoff Library

crude. On the second, Zworykin had obviously filed his application four years earlier than Farnsworth (Stashower, 2002, 216). The court concluded, however, that testimony by and on behalf of Zworykin regarding the date of operation was "not convincing" (Godfrey, 2001, 75). The physicist was too honest to dissemble on what he proposed in 1923 and what he understood later about charge storage, regardless of Westinghouse's efforts to backdate his 1925 insights into the earlier application. Moreover, for reasons known only to the legal counsels for Westinghouse and RCA, the latter company could not draw on Westinghouse's internal reports or the 1925 tube. The history is not clear, but the two corporations had their own conflicts over the rights to Zworykin's television work in East Pittsburgh before 1928. Meanwhile Farnsworth's lawyers used his notebooks to document the operation of his system in 1927 and 1928 using the electrical image that was at the heart of the case.

In July 1935, the examiner ruled for Farnsworth. Zworykin's 1923 application had no right to claim an electrical image as defined in Farnsworth's patent because Zworykin did not originally specify the "discrete globules capable of producing discrete space charges" and therefore did not generate "an electrical image that is scanned to produce the television signals" (Fisher and Fisher, 1996, 236). These were original to his 1925 patent, but RCA had to defend the earlier application because of the patent claims filed by other inventors in the two years between Zworykin's applications. RCA regrouped, rearranged its arguments, and appealed the decision to an appeals board in January 1936, only to be denied in March.

RCA's attempt to claim the concept of electronic video imaging had backfired. In a triumph for lone inventors with good lawyers, Farnsworth gained priority for a basic component of any television system where the image would be moved through vacuum inside a camera tube. The iconoscope did not require this feature, but as the 1930s wore on, it became clear to RCA's technical staff that the next generation of camera tubes would draw on Farnsworth's electronic image and low-velocity scanning.

Sarnoff and RCA could have pursued their case in the civil court system but did not. A version of Zworykin's 1923 application eventually issued to Westinghouse in 1938, and RCA, Farnsworth, and other inventors continued to contest claims to other basic aspects of electronic television, including the storage principle and low-velocity, interlaced, and sawtooth scanning techniques. Despite claims that Farnsworth's company lost licensing opportunities because of the patent interferences, RCA had done him a favor. While there was no prospect of a market or a broadcast standard in the mid-1930s for even a working system, RCA established Farnsworth as the source for several essential television patents. Licensed access to these provided Farnsworth's only profit through 1938: $69,000, against $1,034,000 in expenses.

Compared to Farnsworth's costs, 15 percent of which paid for filing patents and defending and asserting the basic claims therein against other people's patents, by 1940 RCA had spent over $9,250,000 on television. Patent expenses accounted for 23 percent of this sum, and $2,650,000 represented research and development costs. The rest paid for field tests, manufacturing costs, and NBC's program expenses. Yet the return on all of this rode on concepts that were legally if not uniquely Farnsworth's. RCA had to include him in the patent pool it shared with GE and Westinghouse. After lengthy negotiations, led by an ex-RCA patent licensing executive, Farnsworth's company negotiated the first nonexclusive patent exchange agreement with RCA in October 1939. After nine years, Farnsworth had finally gained access to an exclusive club of major American corporations at a fraction of the cost RCA invested in the system.

SYSTEMS ENGINEERING AT RCA, 1931–1937

Television is more than a camera and display. Engineers have to integrate those devices with each other through a transmission system that synchronizes and amplifies the video and audio signals, maintains the frequencies, or channels, on which they are broadcast, and overcomes noise and interference from sources internal to the technology and external to the channel. This system, the "black box" understood only by technical experts, interacts with and is subsumed within the broader systems of commerce, politics, and art. The national government, since it controls the electromagnetic spectrum that envelops its citizens, has to decide where to set the standard for transmission, reception, and quality of television broadcasts. Economically, these technical requirements are tied to the cost of the equipment: the receiver to the home consumer, the studio and transmitter equipment to the radio station investing in the new technology. These costs are tied esthetically to the quality of the programs that justify the system in the first place. Production of television content fell between the sophistication of Hollywood film production and the faster pace and smaller budgets of the radio station. Directors, writers, camera operators, lighting and set designers, makeup and costume specialists, sound engineers, and their staffs all had to work out acceptable levels of quality in their fields for national and local programming.

Developing the system and the data necessary for evaluation is an interactive process. Information from early component and system field tests contribute to the calculations of ideal requirements; further tests result in feedback to the requirements, revealing new approaches, improvements, or limitations that the researchers have to address. There may be bottlenecks in particular subsystems that hold back the overall output, but they can also result in leading-edge technologies that overcome flaws elsewhere in the system. The system should be greater than the sum of its parts as engineers and scientists exploit complementary advantages among components and minimize their flaws.

Throughout the 1930s, under Sarnoff's leadership and with corporate financing, RCA pushed steadily forward with the design, development, and testing of an electronic television system suitable for broadcasts across a continent. Led by Elmer Engstrom, RCA's engineers carried out five field tests between 1931 and 1937. It was an incremental process of increasing knowledge through practice and it ensured that the system RCA would propose to the federal government would operate as expected at an objectively derived level of quality. The investment in these tests established the value of the system, the equipment that RCA would make and sell, and the patents that it would license to the rest of the broadcast and radio industries. What was the goal? The television committee of the RCA-influenced Radio Manufacturers Association (RMA) in 1932 wanted reasonably priced, "quiet and satisfactorily illuminated picture equipment for the home" (Slotten, 1998, 82).

The first test, in 1931–1932, used broadcasts from the new Empire State Building and represented the last gasp of Nipkow television at RCA. For it became apparent that Zworykin would not meet his deadline and that 120-line Nipkow television would not provide the quality of reception desired. One benefit was RCA's discovery that the so-called very high frequency (VHF) region of the electromagnetic spectrum, between 40 and 80 MHz, was acceptable for television broadcasting, despite the interference from motor ignitions. This was significantly higher up the spectrum from the 2 to 3 MHz used in earlier transmissions and it indicated that there was space for high-definition television as well as radio on the airwaves. But, as Engstrom wrote, "an image of 120 lines was not adequate unless the material from film and certainly from studio was carefully prepared and limited in accordance with image resolution and pickup performance of the system." In addition the increasing brightness of the CRT in the receiver resulted in screen flicker that "was considered objectionable" at 24 frames per second (Burns, 1998, 412).

Two years later Engstrom's groups staged two more tests, this time in Camden. The first, early in the year, featured the debut of the iconoscope with its 240-line, 24-fps video. The second involved a wireless relay of video at 44 MHz from the Empire State Building to Arney's Mount, New Jersey, and then 79 MHz to Camden. The engineers put the cameras on themselves, test patterns, cartoons, movies, and outdoor subjects, and concluded that they needed to increase the power of the transmitter, increase the definition of the image, and resolve the annoying flicker that appeared at 24 fps with the ever-brighter kinescopes.

The second issue posed an interesting question: how much definition was necessary for moving images on a television to match the quality of motion picture film? During the 1933 tests, Engstrom carried out a series of experiments with volunteers to answer this based on visual acuity, just noticeable differences in resolution, the size of the display, and the viewer's distance from the display. His results confirmed what others had calculated for televised stills, that around 400 lines offered the equivalent of home movies shot on 16mm film.

During another set of tests in 1934, RCA increased the definition to 343 interlaced lines and the frame rate to 30 per second. Randall Ballard applied for the basic patent in electronic interlaced scanning, in which the electron beams scan odd and even lines in alternate frames. The human eye blends two consecutive frames into a complete image, thanks to the retention of vision. Interlacing requires an odd number of lines and processing is eased by using numbers comprised of odd factors less than ten: thus 343 equals $7 \times 7 \times 7$. Using 60 fields with half a picture, thus reduced to 30 flickerless frames, in this arrangement also avoided interference with the national power grid frequency of 60 Hz because the frame rate is one of its divisors. The company's engineers also developed an electronic impulse generator to handle synchronizing and blanking signals, finally making RCA's system completely electronic.

In May 1935 Sarnoff announced during RCA's annual meeting a million-dollar plan to field-test RCA's system in the New York area. This included construction of a state-of-the-art transmitting station in New York City; manufacture of about 100 receivers for observation of the reception up to 45 miles away in the metropolitan area; and development of programming to test studio broadcast techniques. Two weeks later, RCA began 4MHz field tests from its Empire State Building transmitter. Studio production and motion picture scanning took place at NBC Studio 3H in Rockefeller Center, which was wired to the transmitter through both a coaxial cable and a UHF microwave signal. RCA executives and engineers watched a picture a little more than five by seven inches in area. The tests helped RCA's research staff improve the transmitter and receiver antennas. They also simplified circuitry that reduced the effects of electrical interference and multipath distortion, or signal reflection, off buildings. By the tests' conclusion in 1937, RCA had raised the frame definition to 441 ($3 \times 3 \times 7 \times 7$) lines, and its executives petitioned the FCC for approval of a standard.

SETTING A STANDARD, 1928-1938

The question remained, how high was high enough for the definition of the image? Without a standard answer, no one could invest in any network infrastructure or component production with any certainty, and the household consumer would not buy a receiver that would become obsolescent as the definition increased. RCA had an answer that not everyone in the industry agreed with. Neither Philco nor Zenith, as leading radio manufacturers, nor DuMont, as an aspiring television company, wanted it to monopolize television technology. The correct standard for television was a question to be settled by the U.S. government.

The Federal Radio Commission (FRC) had initially permitted experimental broadcasts using 10-kHz channels in 1927. A year later Charles Jenkins began campaigning for a television standard and commercial licenses. The RMA's Television Standardization Group supported a 48-line, 15-frame system in which scanning of images took place from left to right, top to bottom, in keeping with the clockwise motion of the Nipkow spiral. The RMA also recommended 60-line transmission, which ignored Sanabria's 45-line broadcasts in the Chicago area. All of these formats raised questions about a synchronization standard: was it to be part of the transmitted signal or resolved by the viewer with a remote control?

If the FRC had agreed in 1928 to a standard based on the RMA's assertion that "a large potential audience in the broadcast band is already at hand," then television would have been frozen in a very crude form indeed in the United States (Udelson, 1982, 43). The FRC declined to rule, however, asserting the need for an acceptable public standard, and that bar kept rising with the claims and demonstrations of various inventors and companies. The FRC did, however, retain the left to right, top to bottom scan that is still used today, the legacy of the nineteenth-century graduate student in Berlin.

Based on a recommendation for minimally acceptable video by RCA's laboratory director, Alfred Goldsmith, the FRC began licensing five 100-kHz channels in January 1929 in the shortwave spectrum between 2 and 2.9 MHz. RCA's researchers published the first article on the technical standards necessary for commercial television that September. Because television stations several hundred miles apart and on the same channel still interfered with each other's signals, it became clear that more bandwidth would be necessary for "true television service of permanent interest to the public" (Udelson, 1982, 41). By 1931, the FRC added three experimental VHF channels between 43 and 80 MHz, with bandwidths between 1.8 and 20 MHz.

The FRC, which Congress reorganized as the Federal Communications Commission (FCC) in 1934, continued to encourage the radio industry to agree on a standard before asking for government approval. Late in 1935, after a tour of RCA's, Philco's, and Farnsworth's labs and demonstrations, the FCC concluded that the companies needed to improve their systems further. A year later it opened hearings on the issue at which Sarnoff laid out RCA's position. RCA had already demonstrated its complete system in April 1936, where research supervisor Ralph Beal said that "home television is at least eighteen months away" (Abramson, 1995, 150). Sarnoff, therefore, felt no urgency for commercialization. In his statement, he balanced RCA's interest in the allocations of higher frequencies of the electromagnetic spectrum for experimentation, while the public enjoyed current technology in the lower-frequency regions once manufacturers and the government agreed to a standard.

Some competitors were less certain than Sarnoff that RCA's standards or preferred pace of innovation should set the course for television. Eleven companies, most of which had not worked on television technologies, asked the FCC how far RCA's control of broadcasting "will be allowed to extend into the television field" (Slotten, 1998, 84). The FCC kicked the technical issues back to the RMA for resolution while it evaluated the social factors, most of which boiled down to an affordable system for the average American, and the interests of other industries—the press and Hollywood, as well as radio—in seeing innovation take place slowly, if at all.

RCA had already recommended the RMA involve itself in the development of the new system. In the mid-1930s, seven companies were involved with the technology: RCA, Philco Radio & Television, Farnsworth Television, Allen B. DuMont Laboratories, GE, Don Lee Broadcasting System in Los Angeles, and Zenith Radio Corporation. With the exception of Farnsworth, RCA aided the others by providing information from its research, testing licensees' equipment, free consultations on design issues, and service as a clearinghouse for hundreds of new patents relevant to television.

The RMA's Committee on Television began meeting in April 1936. Members, besides RCA's representatives, included engineers from Farnsworth, Philco, Hazeltine Service Corporation, and Bell Telephone Laboratories. RCA feted the other members with demonstrations and a lavish banquet in July at which Sarnoff praised his competitors at Philco and Farnsworth: listeners "mustn't think that RCA was everything in television" (Stashower, 2002, 221). On the basic question of a broadcast standard, the committee adopted Philco's recommendation for a 6-MHz bandwidth for each television channel, and something more than 300 lines per frame interlaced sixty times a second. A month later, the committee agreed on 440 to 450 scanning lines with a 4:3 aspect ratio for the display, which also happened to be the ratio for 35mm film.

While the broadcasters, manufacturers, and government debated how to broadcast television, RCA began transmitting on the RMA standard up to 90 miles away in June 1937, and demonstrated it to thousands of visitors to NBC in Rockefeller Center. When RCA introduced its 441-line television to the Society of Motion Picture Engineers convention in

October, the screen was far brighter, in black and white rather than the black and green, thanks to Humbolt Leverenz's new phosphors, with a 7-inch by 10-inch display. Sarnoff, who addressed the society, also announced a projection system for theatres, which provided a video image 12 feet square. By 1938, scientists and engineers had improved the sensitivity of the iconoscope by a factor of twelve in six years.

In December 1937, close to the timetable RCA's Beal predicted in 1935, the RMA committee forwarded its standard to the FCC. Most of the membership regarded it as provisional, good for use during continued field experimentation with transmitters and receivers. Yet within that standard, described as flexible enough to permit technical improvements, RMA members continued to disagree. During the spring of 1938, RCA even sided with Farnsworth on one issue, and against it on another. As the group settled some technical issues, other issues arose, and the number of television subcommittees rose over 1938 and 1939, complicating the prospect of agreement.

For Sarnoff, the only obstacle to commercialization now was the development of programs, not more technology. He had already begun to argue this point but, with the economic recession in 1937, did not push the point. Over 1937–1938, however, NBC staged over 250 studio broadcasts; in 1938, RCA and NBC engineers began testing mobile television units in New York and Washington, DC, that could transmit live video feeds on location by microwave dish back to the transmitter.

THE BIRTH OF AMERICAN TELEVISION BROADCASTING, 1939

To prod the FCC among manufacturers and the public, Sarnoff told the members of the Radio Manufacturers Association (RMA) in October, 1938, that RCA and NBC would begin regularly scheduled television broadcasting with the opening of the New York World's Fair in six months. "Television is now technically feasible," he asserted. "The problems confronting this difficult and complicated art can be solved only from operating experience" (Bilby, 1986, 132). The broadcasts would cover only the 50-mile radius around the Empire State Building and take place for two hours a week, without advertising, since NBC would be using its experimental license until the FCC approved a standard. The Columbia Broadcasting System (CBS), which had bought an RCA transmitter and installed it atop the Chrysler Building, was rumored to be considering regular television broadcasts, and Sarnoff urged manufacturers to take advantage of their licensing agreements to draw on RCA's intellectual, technical, and production facilities for making their own receivers.

Was this the best way to resolve the debate? Sarnoff made his decision based on the advice of his technical, manufacturing, broadcast, and marketing executives. If they were ready to produce a system meant for a mass market, so was he. He also knew too well that he had pushed RCA's board to invest ten million dollars in television's innovation over the last twelve years. Finally, he acted on pride, alternately reluctant to admit that some of the opposition might have a point about refining the standard that RCA and the RMA proposed, and furious with resistance based on impractical technologies and corporate self-interest.

Exactly six months later, on a brisk and cloudy afternoon in front of the RCA pavilion at the fairgrounds in Queens, David Sarnoff ignored the raw April weather and read a 7-minute speech to an NBC television camera and microphone. A few chilled engineers and assistants watched nearby, and about 100 people watched Sarnoff speak on RCA's new TRK-12 televisions at the RCA Building in Rockefeller Center. It was actually ten days before the Fair's official opening but Sarnoff always had a strong sense of anniversaries. "Today we are on the eve of launching a new industry based on imagination, on scientific research and accomplishment.... Ten days from now this will be an accomplished fact" (Sarnoff, 1968, 100).

On April 30, President Franklin Roosevelt opened the Fair as the first president on television. For the next eighteen months, thousands of visitors trooped through RCA's exhibits and saw a live television pickup of those behind them. They also saw RCA's $995 receiver, its cabinet crafted from transparent Lucite plastic, with its 12-inch CRT pointed vertically at a mirror on the underside of the cabinet lid. Over the next year, NBC televised 601 hours of programming to its New York audience. RCA, GE, DuMont, Philco, and a couple of other manufacturers rolled out sets in the spring of 1939, while others began planning for production in the next year.

Few consumers bought the receivers at prices between $400–600, however; radios and phonographs sold for $10–250, and one could hear an enormous amount of professional content on either system. By comparison NBC broadcast live day-time events like college sports and fashion shows from department stores, and evening programs featured boxing and ice hockey from Madison Square Garden along with old movies, none of which cost much to produce. In addition, beyond the uncertainty of the FCC's stand on commercial standards, Germany's invasion of Poland on September 1 changed the focus of RCA's manufacturing and research activities. Two days later, Sarnoff ordered RCA production divisions to begin

On a cool and overcast April 20, 1939, RCA president David Sarnoff stood in front of the RCA Pavilion at the New York World's Fair in Queens and announced the beginning of regularly scheduled, electronic, television broadcasts in the United States. Several hundred viewers watched him on 441-line, monochrome, 12-inch displays. Sarnoff's move served to focus the Federal Communications Commission on approving a 525-line broadcast standard in 1941, based on RCA's ten years of innovation. David Sarnoff Library reorganizing to meet the needs of the armed forces. Consequently, RCA Victor's sales division curtailed earlier ambitions for the innovation of television. Instead of 75,000 receivers, it hoped to sell 25,000 in 1940; by the spring of 1941, when RCA stopped making home television sets for the duration of World War II, it had sold not more than 2,500 receivers.

Yet the company continued to encourage other companies to join in, providing portable equipment to the Don Lee Broadcasting System in Los Angeles; demonstration equipment to Westinghouse, Bell Labs, and Stromberg-Carlson; receiver components to Zenith Radio; and CRTs and related equipment to amateur enthusiasts. On October 2, RCA signed a nonexclusive cross-license agreement for patents with Farnsworth Television, which added another ally to RCA's marketing efforts. The sooner consumers saw televisions for sale, the sooner Farnsworth could receive his royalties.

RESISTING A STANDARD, 1940–1941

As the RCA-led rollout languished, however, one industry newspaper called it "Sarnoff's Folly" (Fisher and Fisher, 1996, 289). Meanwhile the standards on which the RMA committees had been working came under new criticism. Over the previous three and a half years, the RMA standards groups agreed unanimously or voted with a strong majority for one approach over another. Philco now recanted its earlier approval on several issues and Du-Mont, which never joined the RMA, claimed that it could raise the number of lines per frame from 441 to 625 and adjust the scanning standard and frame rate frequency across a wide gamut of options. Sarnoff was furious, calling the opponents "scavengers" and "bloated parasites who feasted on the products of RCA research" (Stashower, 2002, 223). In return Zenith ran an ad that showed him as a "Televisionary" King Kong, destroying the radio industry (Bilby, 1986, 132).

The FCC began new hearings on the standards in January 1940. It revisited the RMA's process and focused on the fear that a standard fixed too soon would prevent further technical innovation. While most of the RMA's members favored limited commercial broadcasting with the current standards, Zenith joined Philco and DuMont in opposition. CBS, the other leading broadcaster, could not offer the FCC an opinion on what to do, but its chief executive, William Paley, had little interest in buying television equipment for the network. RCA reported optimistically on the sale of 130 receivers at reduced prices using installment payments, and predicted that manufacturers could sell 25,000 sets in 1940, 60 percent of which would be RCA's.

The commission's continued indecision reflected in part the rotating chairmanship of the 1930s, and now reflected the politics of its new chairman, James Fly. Fly's long experience with the exploitative nature of electric power utilities while working for the Tennessee Valley Authority gave him a deep dislike of monopolies. RCA and Sarnoff's behavior with regard not only to television technology, but to its control of two NBC broadcast networks, reinforced this attitude. Consequently he was only too eager to believe the claims of television entrepreneur Allen DuMont that he could build a receiver compatible with a wide range of frame rates and scanning lines, and then Peter Goldmark's suggestion that the country move straight to color.

On February 29, the FCC issued an Order for the regulation of telecasting without deciding on a standard. The Order reflected what the commission saw as insufficient consensus among the RMA's membership. It declared that "emphasis on the commercial aspects of the operation at the expense of program research is to be avoided," and established two types of television broadcast stations: one for experimental research and the other for experimental programming (Slotten, 1998, 93). The FCC feared that establishment of a standard while permitting expanded programming would result in a frozen technology, with no incentive on the industry's part to continue to improve the quality of the system. It also warned members of the industry—that is, RCA—from trying to establish a de facto standard, but encouraged them to broadcast an undefined but "limited" amount of commercial programming (Udelson, 1982, 148).

Sarnoff met with Fly early in March. He explained RCA's plans for making and selling up to 25,000 receivers in the New York area in conjunction with its limited commercial broadcasting from WNBT in September 1940. After seeing the newspaper ad that RCA would use to promote the sale, Fly complimented Sarnoff on the plan. He was less pleased when the campaign began in March in the *New York Times* and *Herald Tribune*. The FCC promptly eliminated the experimental programming licenses and called for new hearings in April. Its press release accused RCA of doing what the February order warned against, harming the public by intimidating its competition into abandoning innovation at the same time that it threatened to leave large numbers of the public with obsolete sets.

Major newspapers in New York, Philadelphia, and Washington, DC, criticized the overregulation of the FCC and its "alien theory of merchandising." Distributors and merchandisers decried the loss of new jobs as a result of the decision. Fly's categorization of CBS and Philco as the "little fellows" bullied by "Big Business" rang hollow to many observers

(Udelson, 1982, 149). It also drew the attention of Senator Burton Wheeler, who called the principals, in particular Fly and Sarnoff, in for a hearing before the Interstate Commerce Committee.

Fly had already opened the new FCC hearing by emphasizing that it was not intended to cover new ground or the design or marketing of receivers, but whether commercial activity in programming and receiver sales would freeze the "art" of television. It was a curious way to justify a hearing just five weeks after the first one, and he had some difficulty explaining it to the senators. Fly likened television systems to those involving a lock and key. If either is changed, the other is useless, and Fly wanted to give Philco and DuMont the benefit of showing whether their latest innovations would improve the standard before establishing one based on RCA's work of the previous ten years.

Besides RCA, DuMont, Farnsworth, GE, and Zenith had designed receivers for the RMA's 441-line, 30-frame interlaced standard. Philco and DuMont still opposed adopting that as the FCC standard, favoring lower frame rates and higher line densities per frame. For them, the possibility of increased flicker with lower frame rates and the need for a dimmer picture to mask it was offset either by DuMont's claims for its phosphors or by the need for higher line densities for the large picture tubes in development. Witnesses agreed that they could build receivers that could receive two or more standards but disagreed on the additional cost. They also generally agreed that they could improve the quality of reception significantly within the RMA standards.

To show that it was continuing to innovate, RCA demonstrated the state of its current work in television, including microwave relays, antennas, phosphors, picture tubes, its new Orthicon camera tube, and a 15-inch by 20-inch home projection television. Of the 600 engineers and scientists employed in electronic research at RCA, 100 worked on television, half of them full time; 40 more developed and designed commercial equipment. Sarnoff, asked by President Roosevelt to have lunch with Fly, refused: "Mr. President, this problem is not in the stomach but in the head. There's no room for compromise. The public either will or will not be allowed to have television" (Lyons 1966, 220).

THE NATIONAL TELEVISION SYSTEMS COMMITTEE, 1940–1941

To resolve the impasse, Fly asked Walter Baker, a vice president at GE and the RMA's director of engineering, to organize a new standards group,

Acutely aware that the new television industry would need prerecorded content to fill expanding hours of broadcasting, RCA tried to convince Hollywood that television could help the movie business. In the winter and spring of 1941 it allied with Twentieth Century Fox to screen live events using a high-voltage projection TV in the balcony of the New Yorker Theatre on 54th Street in Manhattan. David Sarnoff Library

the National Television Systems Committee (NTSC) that would be more inclusive than the RMA. Baker invited all interested and technically competent parties to join, and documented their discussions with printed minutes of each meeting. Forty-one companies participated in the NTSC's work under nine subcommittees, which began technical and subjective tests on the proposals made in the April hearings. Staff used lab space at RCA, GE, Philco, Hazeltine, and DuMont for eight months to examine signal synchronization, scanning lines and frame frequency, AM and FM sound, color television, and UHF transmitter range.

The NTSC submitted its report to the FCC in January 1941; the major change involved using FM for the sound channel. Color television, which CBS had demonstrated using a three-color filter wheel on a wider ultra-high frequency channel, was tabled but encouraged in further research. It also rejected DuMont's flexible standards. Again, DuMont, Philco, and

Zenith dissented, leading to another hearing in March. Only at the start of this hearing did NTSC consultant Donald Fink recommend increasing the scanning density to 525 lines ($3 \times 7 \times 5 \times 5$) from 441. The increase had been possible since RCA's George Brown invented a vestigial sideband filter in 1938 that nearly doubled the horizontal resolution possible in a given bandwidth. Tests on viewers at Bell Laboratories showed that different line densities represented a trade-off in horizontal resolution, from side to side, and that few people noticed the difference between the two choices—at least on contemporary CRTs.

On May 2, the FCC finally approved commercial television broadcasting, effective July 1, for eighteen 6-MHz channels using 525 lines per frame, 30 interlaced fps, and FM sound. Aside from shifts in channel assignments in the electromagnetic spectrum, the 1941 standard remained the basis for analog television broadcasting in North America for the rest of the century and beyond. Each station was obliged to run at least 15 hours of video a week. Only WNBT in New York committed to that quantity before the war and began telecasting sponsored by, among others, Sunoco gasoline and Bulova watches. The latter's ad, the first, featured a clock face ticking for a minute; NBC charged four dollars for the afternoon spot and eight dollars in the evening.

Yet when the Federal Communications Commission approved the television standard, RCA's attention was elsewhere. Beyond servicing the television receivers already sold so that they could receive the higher number of lines, the company focused its sales efforts on established products. Its telecasts through NBC did not reach beyond the 5,000 receivers in the New York City area, and it engaged in little professional studio production for two reasons. First, it was expensive, and second, it was still extremely hot for performers under the lights needed to pick up a suitable image. Farnsworth's image dissector had required over 94 kilowatts of lighting indoors; early iconoscopes required one-fourth that amount, or about 240 100-watt light bulbs. Hugh Downs, who read the news once for the 400 television owners in Chicago in 1943, recalled that he had once "looked, momentarily, on the face of the sun itself. Never have I felt such sheer withering force of light as I felt during that tormented quarter-hour" (Ritchie, 1994, 94).

The solution to practical cameras lay on the one hand with incremental engineering, and on the other in the radical transformation afforded by a quite different application. RCA's staff had pushed television from non-standard low-definition imagery to the cusp of mass-market quality in just over ten years. But its official birth in 1941 resulted in a stillborn infant. Over the next four years, the interruption in commercial service caused by the American entrance into World War II led to far better cameras and an improved, cheaper, and finally profitable system. We can measure its success by its durability and persistence as a framework for later standards. As we will see in the next chapter, however, that successful life was not predetermined.

4

Working for a Living: Television's Commercialization, 1941–1966

With the approval of the American standard for commercial broadcasting in 1941, many histories of television come to an end. Yet here the story begins for the technology's commercialization, one that earned its keep as the most complex system yet produced for the mass market. Within a generation, television became a ubiquitous part of American and western culture, with receivers plugged into 93 percent of all American households. Meanwhile manufacturers, broadcasters, and governments began developing the formats watched by viewers into the twenty-first century. These included improvements to monochrome television; the expansion of available channels through broadcasts in the ultrahigh frequencies (UHF) of the electromagnetic spectrum; and the innovation of government-approved color television systems.

This period represented the acme of RCA's and American influence on the technology. Under David Sarnoff's leadership, RCA established a virtual monopoly on television patents at the same time that the United States enjoyed a unique intellectual, innovative, and economic preeminence in world affairs. The country benefited from a European "brain drain" before and after World War II; its isolation from the destruction overseas; the isolation of its domestic market; government investment in research and new factories; and a generation that grew up through the Depression and World War II in a collective society, determined to prosper.

By the 1960s, this environment began to change. This chapter covers the application of television to military uses; the commercial battle for the innovation of and between monochrome and color television; and the diffusion of television around the world. In all of these developments, we can see how people who invent technologies work with, and sometimes against, people who make, sell, buy, or regulate things invented to improve the quality of life.

ALL THIS IN WORLD WAR II: USING TELEVISION FOR MILITARY PURPOSES

The five-year gap between the establishment of a broadcast television standard in 1941 and the resumption of receiver production in 1946 raises a question about how wars affect economies. Without World War II and military investment in television, would RCA and other companies have improved the system as effectively? In this case, wartime innovations, their mass production, and engineering experience with very high frequency (VHF) and UHF communications systems probably offset the loss of continuity in commercial broadcasting and incremental improvements. By 1946, as the government eased controls on the production and sale of thousands of items, RCA and other companies drew on government-built factories; government-funded innovation of components including displays and cameras; and employees now experienced in designing and making high-quality electronic components. The result was that television program producers enjoyed more versatility in the production of broadcasts than ever before, and consumers could buy cheaper receivers tuning in more channels and showing images under conditions barely imagined before the war.

Another question comes to mind. What contribution did television make to the war effort that justified the government's investment? Historians have paid little attention to its role in the war, compared to the atomic bomb, radar, and jet-propelled aircraft. The lack of visibility is ironic since TV was so much more advanced and publicized in the 1930s than the other technologies. The ability to see at a distance has obvious applications in guided missiles and bombs, reconnaissance, and communications. By the end of the twentieth century, the United States continued to lead the world in using television for military applications. In the 1940s, however, what was ready for prime time was not ready for wartime.

There were good technological, institutional, and more controversial human decisions responsible for this outcome. The military needed television systems that could operate at commercial resolution despite physical and electronic interference from vibration, wide and extreme temperature variations, and up to 95 percent humidity—all without human adjustment. The variety of possible applications led to a balkanized program, with different organizations within the government and military overseeing the developments by RCA. The Air Force, dominated by leaders devoted to strategic bombing, was primarily interested in finding a way to recycle its "war-weary" heavy bombers; the Navy resisted extensive commitment to drone bombers in part because of a macho culture that valued human skill and bravery in piloting dive bombers. Nonetheless, RCA delivered over 4,500 television systems during the war, and the missions completed pointed the way for future aerial video applications.

ORIGINS

Ernst Alexanderson of General Electric Company (GE) understood television's military value as soon as he began demonstrating his Nipkow-disc system. In January 1927 he wrote Sarnoff, then vice president for RCA. He proposed that an airplane loaded with bombs could have a "radio photograph" camera that would take still pictures to be transmitted to the person remotely guiding the plane (Brittain, 1992, 191). A month later, Sarnoff gave a lecture to the Army War College that expanded on Alexanderson's idea. He proposed "that a radio-television transmitter installed in an aeroplane might be useful in transmitting a direct picture of the enemy's terrain, thus enabling greater accuracy in gunfire" (Bucher, 1952, 34–35). Eighteen months later, Alexanderson and his scriptwriter used military television in the sequel to *The Queen's Messenger* broadcast, late in 1928. In a science fiction drama, a TV-guided missile closes in on New York City, seen as an image that approaches the viewer until an explosion concluded the broadcast. An officer from the British Royal Air Force thought that the United States had revealed television's "possibilities for future wars" (Barnouw, 1990, 63).

None of these ideas represented a practical application. The electromechanical systems were fragile and provided poor resolution. During the 1930s Philo Farnsworth and RCA's Vladimir Zworykin proposed using electronic television systems. Farnsworth's 1933 demonstration to the U.S. Navy led an admiral watching televised smoke rings at the Philadelphia lab to propose that the technology would enable the navy to "scout the enemy with television equipment in a plane, direct the fire of our gunners and make great advances in aerial mapping" (Fisher and Fisher, 1996, 223). Without photoelectric storage, however, Farnsworth's image dissector made even less sense in combat than in the studio, and from 1934 RCA's researchers and engineers undertook the vast majority of work on military television.

This began with Zworykin's iconoscope and his lengthy memo that April to RCA president Sarnoff on a TV-guided missile. Stimulated by reports of the Japanese interest in kamikaze tactics for precision bombing, the scientist proposed a high-tech alternative based on the revolutionary iconoscope camera tube that his team had demonstrated in 1933. Zworykin's "radio-controlled torpedo with an electric eye" would take "the form of a small steep angle glider . . . and equipped with radio controls and an iconoscope camera. One or several such torpedoes can be carried on an airplane. . . . [T]he torpedo can be guided to its target with the short-wave radio control, the operator being able to see the target through the 'eye' of the torpedo as it approaches" (Zworykin, 1946, 294). He estimated the weight of equipment as 140 pounds that, combined with the torpedo's fuselage, would be less than the 300 pounds of explosives that could be directed at a target.

Sarnoff promptly took Zworykin to Washington, DC, to brief the War and Navy Departments. During the Depression, when veterans of World War I camped in the capital for their pensions, the military was not interested in funding the gap between current technology and a prototype weapon. Zworykin wrote another memo late in 1935 on a TV-guided drone missile that a pilot could control beyond the line of sight. RCA's camera and transmitter weighed more than was practical for these uses, but over the next two years corporate and Soviet funding for its television system led to the first demonstration of an aerial TV camera. By the spring of 1939 RCA collaborated with United Air Lines to show that viewers in a Boeing 247 over New York City could watch video transmitted from RCA's National Broadcasting Company (NBC) on the ground. Similarly, viewers in Rockefeller Center's Radio City could watch video from the airplane's camera, thanks to a 700-pound transmitter installed in the airplane with the camera.

MILITARY APPLICATIONS

Before the armed forces could put these cameras into combat, however, engineers had to solve numerous problems not obvious in controlled demonstrations. The challenges of transmitting video from an airplane were far different from those in televising a track meet. Beyond reducing the equipment weight, RCA Victor engineers in Camden, New Jersey, corrected system problems caused by the effects of sharply changing temperatures and lighting conditions, high humidity, and vibration. In addition, the megahertz frequencies that the system used generated multipath transmissions of the video signal. This meant that broadcast video signals from the transmitter arrived at slightly different times at the receiver, whose atmospheric location changed constantly relative to the transmitter, thereby ruining the received picture. To correct the problems caused by a sharply changing environment from take-off to attack, engineers had to invent solutions that kept the camera and transmitter in tune without hands-on adjustments.

The strategic question remained: how exactly would the armed forces put this new technology to work? The technical and tactical possibilities and uncertainties, the different customers, and the alternatives resulted in a vagueness that earned the technology a poor reputation within the National Defense Research Council (NDRC), which was organized to develop advanced technologies for the military. Uncertain, RCA and its clients divided their attention on different projects. Months before Japan's attack on Pearl Harbor in December 1941, RCA was managing eleven of fourteen military TV programs for NDRC and its clients, the Navy and the Army, which controlled the Air Forces until 1947. The armed forces themselves had different priorities. Most of these centered on remote precision bombing of moving or stationary targets, as well as reconnaissance. This raised questions of the delivery format. For the Air Corps, these included glide bombs and older aircraft converted into missiles, as well as gravity bombs guided by a camera in the nose. The Navy also wanted gliders that could carry more explosives and be launched at a greater distance from the target, but it eventually focused on purpose-built drones launched from bombers or from the ground.

THE U.S. NAVY

The Navy started experimenting with radio-guided drone aircraft in 1936 to improve antiaircraft defense training. By April 1939, while Sarnoff initiated NBC's regular telecasts, Lieutenant Commander Delmar Fahrney proposed to his superiors that RCA's TV system be adapted to guided missiles. When Germany invaded Poland that September, Sarnoff authorized funding to build a lightweight TV system. After three months, engineers produced a small iconoscope tube and camera with a 30-pound transmitter that RCA demonstrated for remote telecasting of live news events. Now the military had a lightweight, broadcast-quality system for Zworykin's aerial torpedo. The government approved RCA's

Project Block as an "assault drone" program in 1940. The company soon began delivering Block equipment to the Naval Aircraft Factory in Philadelphia.

Two years later, in April 1942, RCA and the Navy used a TV-guided drone to launch a torpedo at a destroyer in Chesapeake Bay, and another drone to fly into a towed target. The success of this first TV-guided missile led the NDRC to assign the project primary funding. It underwrote RCA's manufacture in Camden of 500 Block I systems for experiments by the NDRC, Navy, and Army on different applications. RCA also trained two other companies to make the equipment and the military began testing TV-guided missiles in southern New Jersey, Eglin Field in Florida, California's Muroc Lake, and Tonapah in Nevada.

The use of military television also gained the approval of the navy's commander-in-chief, Admiral Ernest King. Fahrney and his superior Captain Oscar Smith showed King films of the tests in May. He approved Project Option, granting Smith authority to develop a TV-guided weapon from the prototypes and put it into combat in large numbers as soon as possible.

What happened? Problems with implementing King's order arose in the control over the development process. Smith requested that the Bureau of Aeronautics construct 1,000 drones and 162 control planes. Rear Admiral John Towers, who headed the bureau, resisted the program for two reasons. First, it was one of many wartime demands for new airplanes and skilled personnel, and he assigned priority to production of manned aircraft for the navy's expanding fleet of carriers. Second, he disagreed with the premise that a peacetime prototype should be converted directly into a mass-produced weapon without first testing it in combat.

Towers's resistance extended to his transfer to an influential position in the Pacific Ocean fleet in 1943. His doubts and the success of the American campaigns against the Japanese relegated the drones to limited attacks far from the main offensives. Late in 1944, Fahrney oversaw the launching of 48 drones against Japanese targets in the south Pacific. Distant pilots, steering by joy stick and a 7-inch picture tube, hit half of the targets, including Japanese merchant ships and a lighthouse and radar station at Rabaul naval base.

Beyond the assault drones and tests with the "Glomb," a 300-mile-per-hour glider bomb, the navy also drew on TV for aerial reconaissance. Even as naval leaders fought over the use of guided missiles, RCA and its partners manufactured 4,000 improved Block III TV systems over the last two years of the war. The Air Forces used a sizable number of these for attacks on German positions in occupied Europe.

THE U.S. ARMY AIR FORCES

The American armed forces have traditionally shown little coordination in the development of a military technology. Television was no exception. The NDRC did not establish standard transmission frequencies to control TV-guided missiles until mid-1943. The Air Force had its own programs for integrating the Block equipment into weapon systems at Wright Field in Dayton, Ohio, and Eglin Field. It did not demonstrate televised control of a drone aircraft until October 1943.

We should not be surprised at the differences in naval and air force development of the technology. The bulk of the Air Force's strategy rested on bombing cities and industrial sites, which required an enormous investment in four-engine airplanes. The consequence for television was twofold. First, the Air Force adopted a policy of recycling "war-weary" bombers. In June 1944 General James Doolittle approved Operation Aphrodite. This turned older B-17 bombers into explosive-laden, TV-guided drones that another bomber's crew controlled by joystick and video monitors, receiving images of the drone's instrument panel and from its nose camera.

That summer crews launched two planes and parachuted before the control planes steered the drones and their ten tons of TNT against German V-1 rocket sites around Pas de Calais in France. In September a pilot launched a bomber drone loaded with explosives from England, set radio control, and bailed out, after which another pilot in a following plane guided the drone by camera to a German airfield on Heligoland Island. The Air Force also developed a TV-guided glider carrying one ton of explosive that B-17s could carry within 15 miles of a target. Two Glombs struck German submarine pens at Le Havre, France. Bombardiers directed four more remotely to targets in France and Germany during August. Finally, the Air Force and RCA also experimented with miniature video cameras installed in the noses of 1,000-pound bombs that could be slightly guided from high altitudes. The VB-10, or ROC, after the mythical bird, underwent drop tests late in 1944 and early in 1945, passing evaluation too late for use in the war.

HOW IMPORTANT WAS TELEVISION IN WORLD WAR II?

The answer is, not very much, for the purposes described. On the other hand, RCA systems enabled the remote control of radioactive isotopes at the uranium and plutonium factories in Oak Ridge, Tennessee, and Hanford, Washington state. Pilots in the first American jet plane, the Bell Airacomet, used TV cameras for aerial reconnaissance in 1945. For the postwar atomic bomb tests at Bikini Atoll in 1946, the Navy outfitted reconnaissance drones through RCA's Project RING. This work led to the development of Remotely Piloted Vehicles in the 1950s, a variety of unmanned aerial vehicles used during Vietnam in the 1960s, and the UAVs used by the United States and Israel in the Middle East beginning in the 1990s. The Navy also adapted obsolete Hellcat fighter planes as TV-guided missiles in the Korean War. About ten years later, it commissioned development of the Walleye TV-guided bomb in 1963. The Air Force started a program for its own version in 1967. The use of solid-state electronics and improvements in the imagers and communications links resulted in success rates of 95 percent on targets in the Vietnam War. In the 1990s, cruise missiles guided by satellite and computer programs transmitted video back to headquarters to track their accuracy. Ground-based lasers began to replace television guidance in U.S. munitions by 1995, but these too sometimes used TV cameras for tracking. Lacking the resources of the United States, other air forces around the world continued to use TV-guided explosives into the twenty-first century.

But in the 1940s, the technological reality was that engineers could not master a system spread over hundreds of miles of space and subject to unresolved problems with imaging and transmission in a changing natural environment. The number of aerial flights, or sorties, that used a TV system paled when compared to the number of manned bombing runs by the armed forces. The culture of manned flight and attack, and the need to prove the value of the Air Forces as an independent branch of the armed services, generated little enthusiasm for either Aphrodite or the Glomb. Nonetheless, the experiences with the video and transmission technologies laid a base of skills and interest that the armed forces maintained after the war. It, and other electronic technologies, also provided an industrial base, a skilled set of workers, a group of knowledgeable consumers, and dramatically improved technology.

BUILDING BETTER CAMERAS

All of the efforts at precision bombing relied on the capacity of the camera in the TV camera to provide video images of sufficient definition to identify targets in sharply changing light conditions. The greatest technical legacy of

Group leader Albert Rose, Paul Weimer, and Harold Law examine an image orthicon camera tube in the RCA Laboratories in Princeton, New Jersey. Underwritten by the U.S. military for use in guided missiles during World War II, it improved the sensitivity of cameras by two to three orders of magnitude, demonstrated here by Ray Kell standing under a full moon on the roof of the Labs. The "immy" became the namesake for Louis McManus's Emmy statuette in 1949, and provided all broadcast television imagery until Philips's plumbicon cameras supplanted it in 1964–1965. David Sarnoff Library

military television was the development of the camera tube called the image orthicon, which broadcasters would use for twenty years after hostilities.

Shortly before the war began in Europe in 1939, RCA's Albert Rose took up where Vladimir Zworykin left off with the iconoscope and invented the orthicon camera tube. Where the iconoscope featured an electron beam scanning the same surface on which light from an image generated electrons, the orthicon featured a two-sided target. It also incorporated Philo Farnsworth's low-velocity scanning technique with the iconoscope's thin mosaic of photocells to improve the imager's dynamic range by a factor of five. Thanks to a better understanding of electron optics and the materials involved, Rose's orthicon performed more consistently than the iconoscope under lower light levels, but sudden exposure to bright light generated an overdose of electrons that bloomed and overwhelmed the video image.

The project managers used orthicons only briefly during the war, for RCA's engineers and the women assembling iconoscopes continued to improve them and reduce their cost. With government funding at RCA's new Princeton laboratories, however, Rose, Harold Law, and Paul Weimer transformed the orthicon into the image orthicon in 1943. The world's most complex vacuum tube to that time, the "immy" combined elements of all previous electronic imagers and incorporated an electron multiplier to amplify the video image output.

By separating the steps in transmuting the light energy in an image into electrical energy in a video signal, the image orthicon improved the signal-to-noise ratio over that of the iconoscope by 100 to 1,000 times and eliminated the blooming problem. RCA's new factory in Lancaster, Pennsylvania, began manufacturing the new tube a year later for installation on drone bombers and Glombs. RCA built 250 of them by the end of the war, when the Air Force began applying them to video reconnaissance. RCA's staff also built miniature image orthicons, or MIMOs, which generated video imagery from the front end of the ROC guided bombs.

In many cases the iconoscope outperformed its successor during wartime comparisons. Nevertheless, RCA's technical staff at the Labs and the Lancaster factory understood that the image orthicon represented a significant advance, and that they and the engineers making the camera controls continued to improve it. NBC's engineers, comfortable with iconoscope-based cameras, rebelled. It required a corporate decision early in 1946 to force RCA's network to set the standard of the TV camera industry for the next twenty years.

TELEVISION EARNS A LIVING, 1945–1950

With the war's end in August 1945, researchers and production engineers began converting the accumulated knowledge, experience, and government–surplus factories into the production of cheaper and better television technologies. The U.S. government's investment in military technologies and manufacturing provided multiple stimuli to the creation of a commercial and cultural system. Like the soldiers returning from World War I with training in radio operations, military engineers and technicians experienced in high-frequency electronics and communications by 1945 provided the staff for new television stations and networks, as well as the most eager customers for the new form of home entertainment. With the conclusion of hostilities, the government sold facilities it had paid for, like the specialized vacuum-tube factory RCA operated in Lancaster, Pennsylvania, to the companies that had operated them under contract. All of this expertise, experience, and expanded capacity gained from the war should have enabled Sarnoff and RCA to sell consumers hungry for new forms of home entertainment, and the radio broadcasters, a world of television equipment and programs based on the 1941 standards. Yet it was not that easy. Why?

One reason is that commerce abhors a monopoly. For every system or standard established for one or several companies, another is advocated by a company left out of the process, for whatever reason, be it technical, commercial, or cultural. In this case, one company, CBS, rejected the electronic monochrome standard established in the very high frequencies (VHF) before the war. CBS had followed RCA's lead in 1940 until chief engineer Peter Goldmark saw the Technicolor film *Gone with the Wind*. That led him to revive 1920s electromechanical approaches to color television, which resulted in some impressive demonstrations. One of the objections then lay in the necessity of using three times more VHF bandwidth than monochrome to transmit the three primary colors. The demands of victory in World War II, however, stimulated the research, development, and manufacture of UHF transmitting tubes and circuits for radar and point-to-point communications. Now Goldmark could point to UHF as the solution for broadcasting as well. It offered the promise of an expanded spectrum—one that could accommodate his field-sequential color television system.

The prospect of changing over to an untested system did not appeal to the prewar investors in monochrome television—RCA, Farnsworth, DuMont, and Philco—for technical and commercial reasons. An admitted gadfly, Goldmark was more interested in provoking change than in studying the propagation characteristics of UHF or the manufacturing of UHF transmitters and receivers. As higher frequencies are much more directional and proportionately weaker in signal strength than lower frequencies, it stood to reason that the UHF would pose problems in broadcasting. CBS chairman William Paley had little interest in underwriting technological innovation, and he was not willing to pay for the necessary field tests or production. Those, he and the Federal Communications Commission (FCC) left to RCA, which built the first UHF station and whose engineers tested transmission and made practical UHF equipment.

The second reason was the prospect of losing the prewar investments in the 1941 NTSC standard. With patent portfolios and declining profit margins on radio receivers, RCA and its allies had a vested interest in rolling out monochrome television as a product line. Even without CBS's leapfrogging proposal for color, they faced a challenge. To attract the rest of the radio manufacturing and broadcast industries, television needed more than improved technology. It required a mass market for the equipment and the programming provided through it. In the free market of the American system, this represented the square of a classic chicken and egg dilemma when innovating a system. With two sets of producers—for equipment and programs—and three sets of customers—for transmitters, receivers, and the programs sent through them, neither producer nor consumer would invest unless the other bought into and used the system. Broadcasters had to know and show that an audience existed to watch the advertising that sponsors bought to pay for programs and the equipment to transmit them.

Who would take the risk? Around the rest of the world, governments underwrote the costs for broadcasting, as England, Germany, Russia, and Japan did before and after the war. In the United States, however, the government limited itself to assigning transmission standards and assigning bands of the electromagnetic spectrum for wireless communications. Sarnoff and RCA, with its NBC network, committed to providing programs at a loss while leading the manufacture of equipment for stations and homes. In retrospect, it is surprising that Sarnoff managed to innovate black-and-white television at the pace he and RCA desired, even as they failed to develop color on the same terms.

COMMERCIAL TELEVISION, 1943–1947

At the height of the war in 1943, but after the prospect of allied defeat had passed, RCA's Victor Division and its Home Instruments Department managers began planning for postwar marketing of radios, phonographs, and televisions. Because of the limits on consumer spending during the war and the massive government spending that underwrote salaries for fighting men and working women, all American industries anticipated a surge of purchases with the advent of peace. This was especially true for consumer electronics, where the government had forbidden their manufacture for the duration. With the increased manufacturing capacity threatening to drive down the price of traditional products, the biggest profits lay in turning television from an expensive novelty to an accepted part of a modern living standard. Sarnoff knew this, and he set the company's strategy in December 1944 by telling his executives in the research, factory, service, sales, and broadcast divisions, "The RCA has one priority: television. Whatever resources are needed will be provided. . . . There's a vast market out there, and we're going to capture it before anyone else" (Fisher and Fisher, 1996,

309). RCA had to convince program sponsors, radio station owners, dealers and distributors, and other manufacturers that, in the words of RCA Victor sales manager Thomas Joyce, the public would soon "take home television service just as much for granted as the present generation takes the radio set" (Von Schilling, 2003, 48).

RCA started by reviving broadcasts from its flagship radio and television station, NBC's WNBT in New York City in October 1943. With an eye on the initial male-oriented mass market for televisions in bars and hotel chains, it broadcast a series of sports events live from Madison Square Garden, from rodeo to boxing to basketball. These remote broadcasts had the additional advantage of low production costs compared to studio programs. The audience consisted mainly of wounded servicemen in thirty hospitals in New York City, upstate Schenectady, and Philadelphia, where microwave relays of the broadcast signal provided an early example of network programming.

By the spring of 1944, RCA and NBC executives began spreading the gospel of network television and its advertising potential for product-starved consumers. Allen Du Mont helped organize the Television Broadcasters Association to support the monochrome status quo. The threat to this lay in the prospect that the FCC would revise the pre-war standard to account for the UHF spectrum bandwidth made available by wartime developments.

As the FCC prepared to revise allocations of the electromagnetic spectrum for radio and television after the war, Sarnoff's broadcast and manufacturing rivals rose in opposition to VHF monochrome. E. F. McDonald, president of Zenith Radio Corporation, supported CBS because Zenith was heavily invested in frequency-modulated (FM) radio. Anything that held back television might move consumers to invest in FM radio receivers instead. CBS revived its call for a color television standard. When adapted for use in wider UHF channels, CBS vice president Paul Kesten argued, Goldmark's prewar color system rendered black-and-white television obsolete.

At the same time, however, CBS joined NBC in broadcasting monochrome TV programs. Over the summer of 1944, WCBW in New York began adapting theatrical productions for the small screen as well as a radio quiz show, *Missus Goes A-Shopping*. Promoting television in two incompatible formats, the current monochrome and proposed color, did not help CBS's request to the FCC, which in January 1945 affirmed the prewar VHF standard.

This ruling reset allocations across the spectrum for radio and television applications, while permitting color experimentation in the UHF. The estimated 5,000 owners of television receivers benefited the most from the

FCC decision. Under pressure from manufacturers and broadcasters, as well as labor unions seeking jobs for returning servicemen, the commission kept the prewar television standards and half of the VHF channel assignments. It left open, however, experimentation in the UHF frequencies. Ultimately the American market would need additional channels in each city, and innovation farther up in the spectrum was essential to the "establishment of a truly nationwide and competitive television system" (Von Schilling, 2003, 62).

For RCA and its allies, this was the necessary affirmation of the prewar status quo, and they acted accordingly. Eighty organizations had applications for TV stations before the FCC in 1945, while 24 had licenses. Most broadcasts took place in New York, Chicago, and Los Angeles, and covered sports events and news, including daylong live coverage of V-J Day in Manhattan, celebrating the end of World War II. Neither of these types of programs cost much in terms of scripts, staging, or production compared to studio broadcasts of original or theatrical content.

For CBS, none of this was good news. Chairman William Paley disdained television before the war, seeing it only as a drain on radio revenues. Supporting Goldmark in his prewar color demonstrations was one way to delay the FCC's decision on television standards. During the war, while Paley worked on propaganda broadcasts in Europe, CBS president Paul Kesten invested in Goldmark's advocacy of UHF channels for color broadcasting. When Paley returned to New York in 1945, he found 50 engineers and scientists working on two floors of CBS's building on Madison Avenue. Goldmark argued, and Kesten agreed, that whatever RCA and its allies did with the VHF channels, the public would naturally gravitate to the higher-quality UHF color, with all the advertising revenue that followed. Patents on color and UHF transmission would generate licensing revenue from other manufacturers if CBS persuaded the FCC to approve color television. For an investment of perhaps $2,000,000 in research and development, CBS could reap up to twenty-five times that sum without having to invest in manufacturing.

Goldmark and Kesten combined a superficial technical suavity, the intrinsic appeal of color video, and the prospect of easy income to make a persuasive case. No enthusiast for the expenses of innovation, Paley accepted the investment reluctantly. He did not want to humiliate his executive staff; on the other hand, in the words of executive Frank Stanton, "It was fashionable to be interested in television but not to . . . where it was going to cost you a lot of money" (Bedell Smith, 1990, 276).

In the uncertain regulatory climate, the gamble appeared worth it. Goldmark's team first demonstrated a UHF color broadcast in October

An RCA publicity photo reassured prospective buyers of its new 1946 television receivers of its ease of use and suitability for the parlor. RCA sold 5,400 of the 630TS and its 10-inch display, which its engineers adopted from guided-missile monitors during World War II, for a list price of $375 through installment plans. The design proved practical enough to become branded the "million-proofed" set by 1948, after RCA licensed the design to the rest of the industry. David Sarnoff Library

1945. The FCC rejected the network's petition for color the following month. With no new receivers on the market, however, CBS still had an opportunity to change the commission's opinion without significant political fallout.

Which was why RCA staged a reception during NBC's telecast of a boxing match in the nation's capital in June 1946. NBC analysts projected that it would lose $8,000,000 through 1949 on television broadcasts. Despite Sarnoff's determination, someone had to pay the bills; RCA could not afford further regulatory delays. Until the FCC abandoned the possibility of using UHF to accommodate the incompatible CBS color system, Sarnoff had to ensure that consumers would embrace monochrome TV quickly enough that RCA's 16-year investment would finally net a profit.

The NBC telecast echoed a radio broadcast 25 years before. In July 1921, Sarnoff assisted a network of radio enthusiasts in broadcasting a boxing match between Jack Dempsey and Georges Carpentier to some 300,000 listeners, helping stimulate the consumer boom in radio. Now he presided over an assemblage of 800 politicians and civil servants who watched Joe Louis beat up Billy Conn on prototypes of RCA's forthcoming receivers. Others watched in packed bars or on receivers they had built, while a crowd assembled under the stars at the RCA Laboratories to watch the fight projected on a large screen. The event, sponsored by Gillette Razor Company, proved as successful as the earlier broadcast in the word of mouth it generated from an estimated 140,000 viewers. Thus the question Sarnoff asked the riveted onlookers that night: "Is this acceptable television?" (Von Schilling, 2003, 73) A *Washington Post* reporter enthused, "Television looks good for a 1000-year run" (Barnouw, 1990, 101).

By the following month, the FCC had approved 24 new licenses for TV stations, and NBC began spending $4,000 a week on a 1-hour variety show called *Hourglass*. The format originated on vaudeville stages, and this program established a template that continues to be followed by the network's *Saturday Night*. In the fall, telecast programming passed the weekly totals of 1941; in New York, site of half the country's receivers, NBC, CBS, and DuMont provided hours of afternoon and evening programs—primarily sports events, quiz shows, and talk shows that required minimal scripting, direction, or staging.

Who enjoyed these programs? In a population of 130 million people, well under 1 percent watched on prewar receivers. A year passed before the government relaxed wartime production and price controls; labor unions and industry management resolved differences over wages; and RCA Victor Division's engineers converted wartime receiver designs to use in the home. On September 19, 1946, RCA began production of the first commercial television receivers since 1941. In the last four months of the year factory workers in Camden built 5,400 receivers with 7- and 10-inch screens. The larger set, the 630TS, retailed for $375, which was significantly lower in price—as well as bulk—than the prewar TRK-9 and its 9-inch display. DuMont, Philco, and General Electric joined in, but the numbers did not suffice for holiday shoppers. An annoyed retailer said that RCA "promised me five sets before the end of the year—five and I could sell 500 if I had them" (Von Schilling, 2003, 75).

Meanwhile CBS refined its color cameras, receivers, and transmitters and held hundreds of demonstrations for thousands of people in advertising, in the broadcast and electronics industries, and the federal government. The network's Frank Stanton, a Ph.D. psychologist by training, conducted surveys of audiences to document the desire for color, and the company renewed its petition for a UHF color standard in September 1946. For

Goldmark and Stanton, who succeeded the retiring Kesten, everything rode on the commission's approval. On the advice of an FCC member, Stanton withdrew the network's applications for VHF television stations in four of the richest cities in the country. The FCC had rejected CBS's earlier requests in January and November 1945; if it denied CBS a third time, Paley would pull the plug on the investment and Sarnoff would lead NBC's radio station affiliates into the certainty of the monochrome system that RCA controlled.

As appealing as Goldmark's demonstrations were, RCA criticized them at the FCC hearings on technological as well as commercial grounds. The technical critique was two-fold, in the method of reproducing color and in the use of UHF. John Logie Baird demonstrated a field-sequential color system in 1928, just four years after he showed monochrome TV in England; AT&T followed shortly after with its version in the United States. The technique required synchronizing sets of red, green, and blue filters spinning in front of the camera and the display. Through the persistence of vision that occurs in human perception, high-speed spinning of the filters meant that viewers overlay in their brains successive fields of primary colors. The composite result is an exceptional full-color video image.

Large companies evaluate competitive technologies for future markets. When Goldmark and CBS demonstrated live color television in 1940, using a custom camera tube that RCA provided, RCA engineers under Ray Kell built and demonstrated their own color systems. They resumed experiments with field-sequential systems as the war ended. In December 1945, its Princeton labs staff transmitted and showed color and three-dimensional TV using a color-filtered wheel as well as a high-speed mirror drum. They received a number of patents on parts of the system, which indicated that Goldmark's assumptions of licensing income—on a system where the basic patents had expired—were highly optimistic.

RCA's engineers also proved to their own satisfaction that the system contained a series of flaws that made it a technological dead end. To transmit three times the video information using electric motors and spinning color filters was tantamount to what Sarnoff likened to a return to "the horse and buggy when . . . the self-propelled vehicle is in existence" (Sarnoff, 1968, 136). Even with expanded UHF bandwidth, the necessity of transmitting three separate color frames reduced the resolution compared to the monochrome standard. It proved impractical to offer the same 30 interlaced frames per second as monochrome, and the lower frame rate increased the appearance of flicker in a sequence of frames, if not the breakup of a fast-moving object on screen into red, green, and blue images. To limit flicker and the headaches it stimulated in many viewers, Goldmark's team reduced the brightness of the display, where the inefficiency of the filters blocked up to 90 percent of the light from a single frame. Finally, and more disturbing to the viewer at home, the color wheel scaled up in diameter to nearly triple that of the display, while the motive power needed for this wheel rose ever higher as picture tubes increased in size far beyond Goldmark's initial 5-inch displays. In other words, the superficial appeal of CBS color receivers in 1947 would become more impractical as engineers produced ever larger picture tubes.

RCA also criticized the use of UHF bandwidth. When the FCC authorized UHF research for television, RCA's George Brown conducted field tests in 1946 to research the propagation of UHF video signals. There are three drawbacks to using higher frequencies for broadcasting. The smaller wavelengths of higher frequency signals are more directional than lower frequency signals and therefore do not radiate as effectively around obstacles, like buildings, between the transmitting and receiving antennas. UHF is also more prone to absorption by objects, like tree leaves, and atmospheric moisture, and to multipath interference from the delayed reception of reflected signals. Finally, at higher frequencies, transmitters also require more power to send the signal in all directions and as far as lower-frequency channels. Thus station owners would pay more for a UHF transmitter as effective as one operating on VHF. Brown's group soon found that Goldmark arranged his UHF broadcasts with a direct line of sight between transmitting and receiving antennas to ensure the best reception. While RCA built the first UHF station, in Bridgeport, Connecticut, and eventually sold UHF broadcast equipment, the unpopularity of broadcasting in that part of the spectrum ultimately justified RCA's pessimism about its limitations.

CBS hoped that the FCC would overlook these disadvantages and render a decision in its favor by Christmas 1946, which would help strangle monochrome receiver sales. When RCA announced and demonstrated an all-electronic color alternative in October, however, the FCC had to evaluate that system as well. As Sarnoff and RCA's staff took pains to explain, this and other demos over the next four years were works in progress. This version used UHF channels 14 MHz wide, which contained four separate signals to carry the colors and sound.

How do you broadcast live, electronic, color video to millions of viewers? The major challenge in electronic color is attaining and maintaining registration of the three primary color fields of an image, in transmission and display. Field-sequential systems like CBS's avoided the issue because the color wheels at either end used a single video channel to carry successive fields of red, green, and blue, and relied on the brain's retention of vision to fuse them. An electronic system requires scanning line by line of each frame, as in electronic monochrome systems. For the TV studio where space was less of an issue, RCA's bulky color cameras integrated imagery from three image orthicon tubes with their related circuitry. Each tube collected brightness and color information through a series of dichroic, or two-color, mirrors so that the sensor received only one primary color from a scene.

The real problem was in the home. Ray Kell's systems paralleled the cameras by aligning three cathode-ray tubes (CRTs) with a series of dichroic filters and mirrors. If nothing disturbed this arrangement in the very large cabinet, and if the three video signals came over the air without interference, very good color images appeared on the display. It was hardly practical, but it and the problems with UHF transmission were sufficient for the FCC to reject CBS's petition and defer a decision on color television in March 1947.

The commission's members and engineers also had a number of doubts about the CBS system. They had little confidence that "a mechanical filter at the receiver would be accepted by the public," and anticipated "a number of other systems for transmitting color which offer the possibility of cheaper receivers and narrower band widths that have not been fully explored" (Slotten, 1998, 197). Finally, despite the CBS's willingness to defer on applications for VHF licenses to show its commitment to its UHF system, the FCC was suspicious of a company without a factory to make the televisions and other equipment, "a party who will be only secondarily interested in reducing the system to practicality" (Slotten, 1998, 198).

MONOCHROME TELEVISION TAKES OFF, 1947–1950

The decision gave Sarnoff and RCA the opening they needed to promote monochrome television to radio broadcasters and the public. Those broadcasters who had paid $50,000 for an FCC license and thousands more for equipment were losing money on television and scrambling for inexpensive content. AT&T had only begun to install the coaxial cables and microwave links that enabled stations to pick up NBC's network programs, and local productions in this new art gave prospective consumers little incentive to buy a television. The saving grace was sports, for which RCA developed a portable camera and remote transmitting truck. The drama of athletic contests needed no costly studio production or scripts, and provided content for bars and taverns that could afford the receivers. As one Chicago bar owner explained, "Why not bring hockey or fights here, so guys can see sports and drink at the same time. So I buy dese gadgets. Slump in business stopped" (Kisseloff, 1995, 120).

In order to provide the receivers, Sarnoff and executive vice president Frank Folsom invited all of RCA's patent licensees to the RCA Victor factory in Camden in mid-1947. There they and their production engineers received tours of the production lines and tutorials on television engineering. Folsom and RCA encouraged questions, for the point was that the television was too big for RCA to sell alone. If everyone provided quality equipment in sufficient numbers to meet and fuel demand, everyone would profit in an expanding market. This supplemental information to RCA's licensing bulletins helped Philco, Motorola, and 60 other companies build versions of RCA 630TS receiver, which stayed in production through 1948. RCA also reassured dealers and consumers by training 2,000 technicians and offering an optional service agreement on the new technology. By 1949 every radio manufacturer had joined in the new industry.

RCA also promoted the provision of programming. In September 1947, Sarnoff gave a speech at the annual meeting of NBC's radio station affiliates in Atlantic City. In it, he encouraged station owners reluctant to invest in an FCC television license and all the equipment required to produce and transmit television programs, which cost about six times more than radio programs. The advertising that underwrote radio programs and made it a lucrative business, he predicted, "soon will face keen competition from television" (Sarnoff, 1968, 113). For, "as the television audience increases and programs improve—and both results are sure to be achieved—many listeners are bound to switch from sound broadcast to television programs" (114). He warned the owners not to repeat the mistake of the Victor Talking Machine Company, which thought that radio was "a passing fancy" and whose trademarked dog listening to his master's voice "changed its master" as a result (114). The effect of Sarnoff's call for progressive innovation was immediate. Attendees remembered station owners rushing to telephones, and RCA reprinted the speech in its sales magazine, *Broadcast News*. NBC encouraged those who were still reluctant by threatening to cut their radio network affiliation.

A month later, the New York Yankees and Brooklyn Dodgers played an epic World Series that ran the full seven games. It attracted an estimated three million viewers from New York to Washington, DC, mostly in bars, to the sixth game. Ford Motor Company and Gillette Razor Company shared the sponsorship for $65,000 after the baseball commissioner rejected Ford's offer to buy the next ten years for a million dollars.

The number of viewers and rising receiver sales helped lead Swift & Company and its processed meats to become the first network sponsor of a nonsports program, the *Swift Home Service Club*. An NBC vice president expected to see "this program in Hollywood as well as on the East Coast" by 1949 (Von Schilling, 2003, 98). In January 1948, the struggling Du Mont began relaying Ted Mack's *Original Amateur Hour* and its "Wheel of Fortune" from New York to stations in Philadelphia, Baltimore, and Washington. Its popularity attracted affiliations with other stations that couldn't show it on Sunday nights, so Du Mont's engineers began filming the program from a studio monitor. The films, or "kinescopes" as they became known, were copied and shipped for broadcasts later in the week. Now available in digital formats, kinescoped programs started a trend in time-shifting and editing of live programming.

Over two years, Americans bought televisions in astonishing numbers. They watched programs mainly sponsored by cigarette companies in an increasing number of urban areas as the FCC granted station licenses, AT&T connected them through its coaxial and microwave networks across the Midwest, and the number of TV stations quadrupled from twelve to 51. That number nearly doubled in 1950. The mounting number of stations squeezed the supply of VHF channels between 2 and 13, especially after the FCC assigned Channel 1 to other radio services in May 1948. Antenna engineers earned their keep by shaping the broadcast signal from their towering arrays to complement the pattern of signals from other stations.

Yet growing numbers of viewers complained of cochannel interference from two stations with the same channel assignment in the electromagnetic spectrum. In Princeton, RCA engineers Alda Bedford and Gordon Fredendall found they received NBC's Channel 4 from Washington, DC, as well as New York. They solved the conflict through the now-common technique of offsetting one of the channels by a tiny fraction of its assigned frequency, but not before the FCC decided to freeze further applications for three and a half years in September 1948.

In that time the FCC held a contentious series of hearings where the commission tried to balance the interests of established stations and networks with the desire of newer groups for VHF allocations. In April 1952 the FCC approved new allocations that blended 70 UHF channels for new stations with the existing VHF stations. Despite the hope that UHF allocations would increase the competition and coverage in each urban area, the FCC ignored the propagation problems evident in RCA's field tests, the reality that it cost more to operate UHF stations for a given audience than VHF, and that it was more expensive to build UHF reception and tuning into a home receiver.

Despite the technical problems and regulatory controversies, Americans bought five times as many receivers in 1948 as they did the year before. Two programs drove another tripling of purchases, to nearly three million, in 1950. One major attraction, as with radio in 1924, was the political conventions for the Democratic and Republican presidential candidates. The four networks, NBC, CBS, the American Broadcasting Company (ABC) and DuMont pooled their live coverage, which had the benefit of reducing production costs. Delegates and politicians began learning the importance of appearance on-screen before a remote audience and the pivoting cameras arranged around Philadelphia's Memorial Hall.

The other program was Milton Berle's *Texaco Star Theatre*. Berle, a veteran of vaudeville, radio, and film had gained little notice in those media. He found an enthusiastic audience for his physical comedy and well-rehearsed routines with guest stars while hosting the variety show on NBC, which the *New York Times* called "television's first smash" (Von Schilling, 2003, 103). NBC's success in pulling people off the streets and out of bars and theatres on Tuesday nights drove CBS to start what became *The Ed Sullivan Show*. Berle's professionalism set a standard for future network productions; Sullivan's showcase introduced not only Elvis Presley and the Beatles to the nation's viewers but exposed Americans to a multiracial array of pop culture performers until 1971.

COLOR DEVELOPMENTS, 1947–1950

When the FCC chairman resigned late in 1947 for a post at RCA, Wayne Coy succeeded him. Formerly involved in the anticorporate politics of the New Deal and the FM radio group that fought RCA over spectrum allocations, Coy became a partisan supporter of the CBS color system. He disliked RCA's monopolistic position in broadcast technologies and regarded its slowness to innovate color television as arrogant and greedy. In addition, the intrinsic appeal of a simple system that provided color, even on a small screen, contrasted with the uncertain complexities of RCA's promises for an electronic, scalable version. By 1948 the irrepressible Peter Goldmark had squeezed his color system into a 6-MHz VHF channel by reducing the number of scanning lines from 525 to 405. He gained new publicity with a series of closed-circuit transmissions of surgical operations, where the bright lighting, slow hand movements, and dim viewing rooms showed off field-sequential color to its best advantage. The public spirit evident in this work, backed by CBS's position as a broadcaster without an interest in making and selling color receivers, helped persuade Coy that the FCC should revisit the issue of color television. On July 11, 1949, it called for proposals for color television that would fit in the current 6-MHz channels—and be compatible with monochrome receivers—to be submitted in six weeks, with hearings in Washington to begin a month later.

RCA had not focused its research efforts on electronic color since the 1947 decision. Instead its staff had investigated UHF as a possible medium for expanded numbers of television channels. Nonetheless, with the prospects for UHF in decline, the technical staff began mulling a VHF color system that would be compatible with the current monochrome system. In this proposed format, color receivers would show monochrome video, and monochrome sets would receive video transmitted in color, but display it in black and white.

Together these techniques represented a system for transmitting color video electronically within the current VHF channels. Before July 1949, however, RCA's technical staff had not integrated or tested them. Nonetheless, at an emergency meeting in Princeton with RCA's executives about the coming hearings, Dr. George Brown argued that the Labs could demonstrate a monochrome-compatible, electronic color system. He was given full authority to lead the program, and ten weeks to prepare RCA's demonstrations. Two major challenges presented themselves: compressing the color signal into a channel intended for monochrome television, and making a receiver that displayed color as well as black-and-white programs through a single CRT.

The solution to compression came in steps, beginning with Alda Bedford's principle of mixed highs. He recalled matching color threads for his mother as a child in Kansas, and the fact that one cannot distinguish color detail at a distance. In a color video signal, that meant the blue, or high-frequency, details could be transmitted in monochrome. Meanwhile, Clarence Hansell and RCA's research group at Rocky Point, Long Island, had developed during the war a multiplexing system to combine multiple wireless messages into one composite signal. At the receiver, electronic equipment decomposed the signal into its constituent messages, thereby saving the bandwidth in transmission. To apply this approach to the red, green, and blue video components, four engineers transferred from Long Island to Princeton and never returned. Randall Ballard, who invented the interlaced scanning technique for monochrome television in the 1930s, developed a "dot interlace" system for color. This meant that color information was sampled, transmitted, and displayed at each element in a frame rather than at each scanning line, as in another proposed system, or each field, as in the CBS approach. Finally, Bedford stabilized the synchronization of color signals at the camera and receiver by inserting a burst of eight sine waves on the "rear porch" of the signal.

The hearings began on September 26 while Brown and his staff were still testing cameras, transmitters, displays, and other equipment at NBC's WBNW station at the Wardman Park Hotel and locations in the Washington area. Two weeks later, RCA began a disastrous set of demonstrations during a heat wave. Without air conditioning, equipment overheated and distorted the signals so that the primary colors didn't register, or even appear. Brown called the studio at one point "to see whether the transmission was in color or black and white." After the first day, Goldmark testified in response to an FCC member's question, "I don't think the RCA system should be field tested because I don't think the field tests will improve the system fundamentally" (Brown, 1982, 160, 161).

Over 1,500 people from the U.S. government, the electronics industry, the press, and the public generally agreed after watching what amounted to a laboratory test bed on display. "The monkeys were green, the bananas were blue, everyone had a good laugh," recalled Sarnoff five years later (Fisher and Fisher, 1996, 317). Yet there were enough instances when RCA's system worked as intended, leading Senator Edwin Johnson of Colorado to opine that "RCA's system has a potential of acceptance that is comparable to none" (Brown, 1982, 164).

The challenge of registering three colors from three picture tubes onto a single screen contributed not only to the difficulties in demonstration but to a color TV receiver that was the size of a refrigerator. This did not compare well with the CBS receiver, which only looked impractical when Allen DuMont scaled up Goldmark's color wheel and motor to accommodate DuMont's 30-inch CRT. Even RCA corporate staff joined the accepted wisdom that making a color picture tube was technically and commercially impossible. Thus, when Princeton Labs director Elmer Engstrom promised in October that RCA would demonstrate such a device in six months, few gave him much credence.

Engstrom, however, spoke from knowledge of a massive engineering effort that had already begun. In mid-September, he, Sarnoff, and other executives met at the Princeton labs where they assigned Edward Herold the task of making color picture tube. Herold's authority ran across corporate lines of command at the labs and other RCA divisions; he had a blank check to underwrite the experimentation and development.

The project was exceptionally rare in commercial enterprise, similar if smaller in scale to the Manhattan Project to invent, build, and demonstrate an atomic bomb. Various inventors had proposed methods of showing color in a single CRT, and John Logie Baird demonstrated a CRT that showed much of the spectrum through two primary colors in 1944. But none of these were practical for the ultimate goal of mass production. The Princeton Labs' technical staff in television, computing, vacuum tubes, and cameras offered nineteen concepts in the fall of 1949, out of which Herold and a small group chose five for development and testing.

The winner was Alfred Schroeder's shadow-mask CRT. Schroeder had been fascinated with electronic television since he was a student at the Massachusetts Institute of Technology in the early 1930s. He took up the challenge of electronic color during the war at RCA Labs as one of three conscientious objectors working on nonmilitary projects. Late in 1946 he finally gained permission to build a single CRT that would register video signals containing the three primary colors in one image. His "three-tailed monster" contained three necks, each containing an electron gun. At their junction they ran through a single electromagnetic yoke that deflected the electron beams to phosphors coating the face of the tube. The emitted white light passed through red, green, and blue filters taped to the glass and a series of dichroic mirrors that registered the three images on a screen. The result was good enough to invite Sarnoff to Princeton for a demonstration. After he watched a series of color slides transmitted through the tube, he slapped his thigh and said, "This is it—let's go" (Webb, 2005, 86).

Having proved that one yoke could control three beams, Schroeder turned to a display that did not rely on projection. His solution used the common yoke and a perforated screen or mask that was aligned with strips or dot triads of phosphors. Each of the strips or dots glowed red, green, or blue when struck by the beam modulated with that color's video signal. The mask prevented the other beams from hitting the wrong phosphors, millions of times a second.

Schroeder filed his application in February 1947. Partly because a 1943 German patent by Werner Flechsig interfered with, or overlapped, its claims, the Patent Office did not approve it until 1952. In the fall of 1949, however, RCA's Harold Law set about reducing it to practice. The challenge was to align the mask holes with the phosphor triads, which Law solved by building a "lighthouse." This device used three light sources as analogs for the electron beams, which exposed a photo resist at three different angles, enabling the sequential deposition of the red, green, and blue phosphors in the correct positions. By December, Law demonstrated a crude shadow-mask picture tube. He and hundreds of engineers, scientists, technicians, and shop workers spent the next three months on nonstop engineering at RCA's laboratories and factories in Camden and Harrison, New Jersey, and Lancaster to make prototype tubes and receivers.

RCA Laboratories staff took three years to replace Ray Kell's Trinoscope with the shadow-mask color television tube fit inside the same size cabinet as a monochrome display. Alfred Schroeder's "three-necked monster," lower right, showed in 1947 that one electromagnetic yoke could deflect three electron beams. This eliminated the need to align three cathode-ray tubes and color filters to project the video signal. David Sarnoff Library

On March 23, in Washington, DC, a month ahead of Engstrom's promise, RCA demonstrated the results of its staff's labors to the FCC. The 12-inch shadow-mask CRT and its electronic circuitry offered color images from a cabinet designed for a 16-inch monochrome tube. The shadow mask blocked about 90 percent of the electrons from hitting the phosphor, so the picture was as dim as the CBS displays, but it offered higher resolution, and compatibility with the more than three million receivers in American households. Over the next two weeks, hundreds of people who had scorned the company's system in October changed their minds about its quality and potential.

RCA's lab-based technology was still sensitive to electronic interference, within or outside the system, however, and prone to color "dot crawl" and moiré, or interference, patterns in detailed images. Shortly after the shadow-mask demonstrations, Bernard Loughlin at Hazeltine Laboratories on Long Island, New York, solved these problems, to RCA's embarrassment.

Since it cross-licensed broadcast technology patents exclusively with RCA, Hazeltine used its relative seclusion and access to RCA's technical reports to improve on the latter's approach. Instead of treating the color signal as pulses based on the dot-sequential sampling, Loughlin analyzed it as a subcarrier sine wave that was added to the brightness signal. Because of the many mathematical tools developed to work with sine waves in broadcast circuitry, this enabled some conceptual breakthroughs in the treatment of the signal. Loughlin's team now "shunted" the brightness signal through the entire transmission, eliminating the interference between it and the color signal. In addition, they proportioned the color signal to the brightness of the three phosphors. In the presence of interference, this gives the impression of a stable image since the eye is more sensitive to changes in brightness than in hue and color.

COLOR TELEVISION, 1950–1953

By the end of 1950, RCA had integrated Hazeltine's techniques; the television industry's second National Television Systems Committee (NTSC) had started supporting RCA's monochrome-compatible system; and Americans were closing out the purchase of another seven million monochrome receivers. But the FCC pressed ahead with its choice of the CBS system for the nation's color standard. Why? In its own words, RCA's system featured a "prominent dot structure and a marked loss of contrast. Moreover the colors are not true . . . it is obvious that no serious consideration could be given to a system that failed to present true colors" (Abramson, 2003, 44).

On the other hand, there was no practical way to adapt millions of monochrome receivers to accept CBS's color broadcasts. Since Coy and the commission could have deferred on a standard rather than award it to CBS, there were a number of other reasons, all of which contributed to a major embarrassment for the government. To begin with, RCA tried to convince the commission to trust its judgment on three arguments. First, it wanted the government to deny the adoption of the CBS system and avoid setting any standard. That should be left to RCA and its allies. This in turn demanded confidence in the ability of a monopoly reaping huge profits in monochrome television to reduce the demonstrated complexities of compatible electronic color to household technology in short order.

It also helps to see the FCC as a regulatory agency full of political appointments—positions that one lawyer later said "nobody would take unless they wanted to use it as a springboard to private practice afterwards with clients from the industry" (Kisseloff, 1995, 553). In Coy's case, he left in 1952 to become president of two radio-television stations. Such appointments result not only in people overseeing an industry that may pay their future salaries, but also in technically illiterate officials making decisions involving high technology. Regardless of RCA's advances or the growth of the monochrome market, CBS appeared to offer the very attractive prospect of color TV now, through a simple and inexpensive technology. It was similar to the argument that General Electric's Ernst Alexanderson made for Nipkow disc-television a generation before. In 1950, however, the color display was not two inches square but about the same size as that in most monochrome receivers.

The result was that the FCC tried to force a reluctant industry into making CBS-compatible monochrome sets, and approved the CBS field-sequential system in October 1950 as the color standard for the United States. Sarnoff promptly ordered a lawsuit to overturn the decision. After the Supreme Court ruled that the FCC acted legitimately, without evaluating "the wisdom of the decision," CBS began supporting its own standard (Fisher and Fisher, 1996, 319). On June 25 its New York station televised in color a "gala premiere" of CBS celebrities and the New York City Ballet that several hundred people associated with the network watched on 25 field-sequential receivers. What *Fortune* magazine called "a championship dive with no water in the pool," a CBS executive called "a Pyrrhic victory if there ever was one. Another victory like this could ruin us" (quoted in Smith, 1990, 284; Brown, 1982, 214).

After all, CBS was in the business of broadcast entertainment, not manufacturing. Every hour that it transmitted its color programs was an hour where none of the millions of receivers in the New York or Boston areas could display it. CBS lost sponsored advertising for those broadcasts and the ones following in monochrome, and had to underwrite the cost of television production, burdened by additional costs for color.

It is doubtful that Paley sought this outcome from the start. His sponsorship of Goldmark's renewed system in early 1949 represented an effort to push his rival Sarnoff and RCA into innovating a color system from which CBS advertising could benefit. In the fall of 1950, however, Sears, Roebuck and Co., the nation's largest retail and catalog store, agreed to buy field-sequential color TVs from its two suppliers of televisions. It began touting Silvertone color TVs in its catalogs and demonstrated one to local acclaim at its Brooklyn store in July 1951.

With Sears's backing and Goldmark's blithe assurances that his color receivers could be made compatible with the monochrome standard, Paley and Stanton began to believe that CBS could succeed. Over the winter and spring of 1951, they explored purchasing the struggling DuMont company to reap profits from sales, and then settled on Hytron Radio and Electronics Corporation. Based in Salem, Massachusetts, Hytron was a mediocre manufacturer of vacuum tubes and other components, but its Air King television subsidiary on Long Island was one of Sears's TV manufacturers. In June 1951 Hytron's owners became one-quarter partners in CBS, owning $18 million of CBS stock. Their Air King engineers found themselves trying to deliver a pilot run of 100 field-sequential receivers, to be followed by 1,000 more for Sears's Silvertone brand.

The transition from "reduction to practice" in the laboratory to mass production proved CBS's undoing. Goldmark, who had dodged the very real technical problems of his system since 1940, now blamed production engineers for "childish bellyaching" over the challenges of making an affordable, monochrome-compatible, color set (Jacobson, 2006, 2). Because the color wheel had expanded from 8 to 30 inches as Goldmark used larger picture tubes over ten years, the motor to drive it and the magnetic field generated by the motor also gained in size and strength. In the FCC demos Goldmark concealed the bulk of the motor and shielding that kept its magnetic field from interfering with the magnets deflecting the CRT's electron beam. That would not work in the home. Over the summer, the CBS impresario argued with Sava Jacobson, Air King's chief engineer, over the balance between quality, practicality, and cost. To compound the problems, the Korean War made it impossible to assure delivery of the necessary electronic parts.

Meanwhile, RCA and NBC regularly publicized their experimental, compatible, and improving electronic color broadcasts. By the end of the summer, CBS faced deep losses in its core business of broadcast advertising and a $100 subsidy on each of the one million color receivers that Air King had finally begun to produce at $500 apiece. To Paley's surprise and dismay, the receivers did not reflect the quality he expected his network to symbolize. Sears changed its mind about supporting color TV and canceled its order, leaving CBS to try to market its receivers through New York City department stores. Fortunately, the heavy use of mu metal for magnetic shielding inside the receivers provided an excuse to request the suspension of color receiver production by the National Production Authority in November. The United States was immersed in the Korean War and mu metal was vital to various military electronic technologies. At Air King, the executives offered a toast to people looking back at the CBS color wheel "with the same fond reverence now accorded the Stanley steamer" (Jacobson, 2006, 2).

With that, CBS ended its association with Goldmark's system and ultimately with television manufacturing in 1956 after spending close to $50 million on the venture. Meanwhile, Walter Baker of the Radio Television Manufacturers Association recruited 29 organizations and dozens of engineers to the second NTSC's thirteen study panels. For eighteen months they performed field tests and tested further improvements to color signal transmissions made primarily by RCA's engineers. In May 1953, Goldmark seconded the motion to submit the electronic color standard proposal to the FCC. By then Coy and several other members had stepped down, and the new commission approved the standard on December 17.

RCA AND COLOR TELEVISION, 1954–1964

The new standard led to various firsts in color broadcasting, most of which are meaningless because virtually no one outside the broadcast engineers, laboratories, or executive suites could watch the programs in color. Admiral Television Corporation put the first receivers on the market for $1,175 on December 30, 1953, so that it could take a deductible tax loss on its color factory investment. On New Year's Day, 1954, NBC broadcast the Rose Bowl Parade in Pasadena, California, across a 21-station network and RCA arranged with other manufacturers to install color and monochrome receivers in cities across the country. Ten-year-old Edward Reitan was awestruck by the "12-inch screens, so tiny and blurry that you had to look at the larger black and white screens to recognize detail. But it was color and it was gorgeous!" (Reitan) RCA's first production receiver, the CT-100, went on sale at the end of March for $995, about the cost of a used car and five times the cost of a comparable black-and-white set.

The electronic innovation forced by CBS, and RCA's insistence on monochrome compatibility meant that sales of the new color receivers were only slightly better than those for CBS. The color receivers were not nearly as good in mass production as in the field-tested prototypes, and the monochrome signal on them was inferior to that on a monochrome set because the dichroic lenses in the cameras reduced the brightness information available. Contrary to expectations, very few affluent early adopters bought the receivers made and sold by RCA and a host of its licensees. Instead the RCA Service Company staff ran ragged trying to keep the few receivers operating as advertised, leading General Electric Company's president, Ralph Cordiner, to complain that "you've almost got to have an engineer living in the house" (Fisher and Fisher, 1996, 328).

Meanwhile, the public bought monochrome televisions in a boom that rivaled the Internet in the late 1990s, or radio in the 1920s. When the

FCC lifted the freeze on television station licenses in 1952, radio stations and other investors jumped in because the public showed an enormous appetite for programs—and the advertising that underwrote them—from the stations already on the air. AT&T completed its network of coaxial cables and microwave antennas connecting television stations from Boston to Los Angeles in 1951. That September, Harry Truman became the first president to be televised live before a national audience, while NBC began telecasting its *Comedy Hour* live from Hollywood. The west coast connection stimulated the migration of program production from New York and Chicago to Hollywood. As people watched more television, the film business suffered. But the production of TV programs compensated the studios and independent producers handsomely. Within two years, programs made on the west coast dominated the content people watched on the small screen.

Networks staged well-rehearsed monochrome dramas with only an occasional NBC spectacular as incentive to buy color. By 1959, twenty years after David Sarnoff announced regular television broadcasts in New York, 86 percent of all American households had a monochrome television receiver, and left them on an average of five hours a day. Commercial sponsors underwrote broadcasts at a cost of a billion dollars a year. As a result, a new station owner like WTVE's in Elmira, New York, could promise to be "in the black in thirty days" (Barnouw, 1970, 65).

This situation resulted in color sales so low that the Electronics Industry Association refused to publicize them until 1959. Instead of the 75,000 receivers Sarnoff predicted for 1954, RCA sold 5,400 at a loss and then spent more money responding to complaints about poor color and monochrome reception. When sales of color TVs peaked in 1956 well below expectations, every other maker dropped out, leading *Time* magazine to judge that color TV was "the most resounding industrial flop of 1956" (Fisher and Fisher, 1996, 328). Sarnoff, RCA, and NBC were left to promote and underwrite the system alone in 1957. Sarnoff forced NBC's reluctant executives to eat the losses as the sole color broadcaster—outside sporadic colorcasts by CBS—and steadily increase the hours of programming.

By the beginning of the 1960s, however, sales began to pick up. With the market for monochrome receivers nearly saturated, consumers turned to the $495 color sets, which now cost only three times as much. NBC broadcast in full color for the prime-time evening hours, and then for all TV programs. The tipping point may have come in 1960 when it contracted to televise *Walt Disney's Wonderful World of Color* on Sunday evenings, giving families a program nearly as popular as Milton Berle's variety show had been for monochrome in 1948. As importantly, RCA signed a consent decree with the U.S. Justice Department for antitrust violations in broadcast technologies in 1957. As a result, the company agreed to freely license its 3,000 color television and other broadcast patents to American companies, thus removing a long-standing complaint of monopoly by rivals Zenith and Philco, who began making and marketing color TV.

In 1962, eight years after commercialization, the RCA Victor Division finally broke even on the manufacture of color televisions. Two years later, RCA shareholders enjoyed a three-for-one stock split as twenty other companies negotiated to buy color picture tubes from the sole supplier. ABC and CBS finally began regular broadcasting in color during the evening prime-time hours in 1965, when the gross value of color receivers sold exceeded that for monochrome. It would be difficult to argue that Goldmark and CBS had accelerated the pace of innovation. In 1974, twenty years after the FCC's approval of the NTSC color standard, only two out of three households had color TVs.

Nonetheless, from the 1950s to the 1980s, Americans watched a television system defined by a regulated market and the limited channels available in the VHF spectrum. Three broadcast networks—ABC, CBS, and NBC—created an oligopoly that dominated programming. With television's saturation of the country's prospering middle-class homes, network programmers, ad agencies, and corporate sponsors filled people's leisure time with a homogeneous routine of children's shows, cartoons, soap operas, game shows, sitcoms, crime dramas, westerns, and sports championships. Whenever one network led the ratings for viewers with a particular program, the others copied it in the quest for advertisers. The profits to be made from capturing the largest number of eyeballs drove programmers to appeal to the lowest common denominator on the one hand, and to avoid social, political, or economic controversies on the other. With rare commercial exceptions and the broadcasts of the National Educational Television and Public Broadcasting Service, only the evening news shows and live coverage of historic crises compensated viewers looking for more engagement. As early as 1961 the chairman of the FCC, Newton Minow, decried the mediocrity of programming by telling a shocked National Association of Broadcasters that its members profited from "a vast wasteland" (Kisseloff, 1995, 507). The popularity of television as a medium for commercialism, sex, and violence stimulated endless debates on its effect on children and the culture as a whole. None of it offered resolution on the consequences of exposure, especially when researchers compared the American experience to that in other countries.

OVERSEAS

The experience of color television broadcasting elsewhere differed considerably from the United States. Generally, governments not only regulated the airwaves but owned television stations. Each country provided one channel based on transmission standards developed by Great Britain, France, Germany, or the Soviet Union. In keeping with the informal empire of the United States' Monroe Doctrine, most Central and South American countries adopted NTSC standards. In the noncommunist countries in east Asia, over which the United States also exerted political and economic influence, the NTSC standard also prevailed. Japan enabled commercial monochrome broadcasting as well as government-sponsored channels in 1951. Nine years later it began broadcasting NTSC color television, impressing Sarnoff and the RCA engineers who visited that year.

Europe was a different matter. Geographically united and technologically advanced, it had much to gain from a single system, but it was divided nationally, politically, and in monochrome standards. The obvious difference lay in the number of lines per frame. After World War II, The British revived their prewar 405-line system using a five-megacycle channel; the French instituted an 819-line system using fourteen megacycles. Germany and most of the rest of Europe used 625 lines in 7- or 8-megacycle channels. All of these represented efforts to revive national industries that provided employment, protected markets in strategic industries, and promoted national pride.

As national companies began to saturate their domestic markets in the 1950s, and opportunities to exchange programming grew across borders, politicians and industrialists agreed that one standard would enable increased continental and international trade in equipment and programming. The NTSC color standard offered a fresh start. The process of transmitting color became embroiled, however, in French nationalism and the determination to set terms by which the country could maintain a television industry.

European engineers and manufacturers followed RCA's innovation and began reviewing and discussing the NTSC standard in 1955. A year later the Europeans visited NBC and the David Sarnoff Research Center in Princeton to see the system in action. A natural follow-on for RCA and the United States in the postwar era was to make the NTSC color broadcast standard a global one. Not only would this ease the international exchange of programming and technology, it would also keep patent licensing income flowing back to the United States. Over the next ten years, RCA worked with the U.S. government to convince western and eastern Europeans of the benefits of adopting an established system. The flaws in American commercial broadcasts and receivers, however, especially with the need to adjust the color tint on-screen, led to the description of NTSC as "Never Twice the Same Color" and bolstered arguments for European improvements.

French and German researchers invented two solutions that were compatible with monochrome broadcast systems provided by their respective governments: Séquentiel Couleur à Mémoire (SECAM) and Phase Alternating Line (PAL). Both of these represented variations of the NTSC system, where RCA's and Hazeltine's patents continued to comprise 95 percent of the technology. The French, under Henri de France at Compagnie Francaise de Television, developed SECAM between 1956 and 1959. Telefunken engineer Walter Bruch provided the German alternative, PAL, in 1962.

SECAM's improvement consisted of correcting color errors developed in the new magnetic videotape recorders and in microwave transmission of the signal. Both of these became irrelevant as other engineers improved those technologies. Difficult to convert to PAL or NTSC, SECAM made adding effects to the video signal in television stations initially impossible, and added more expense to receivers. PAL reversed the phase of the color subcarrier on alternate lines to cancel flaws in transmission. The NTSC had experimented with the technique at the color field level, and RCA and Hazeltine had rejected Bruch's technique in 1951 for technical and economic reasons. While PAL offered some improvements, it too added significantly to the cost of receivers.

The European Broadcast Union conducted tests of the three systems that showed that no one of them was superior to the other two under all conditions. As a result, the decision became a political one, based on national and economic interests. As the rest of western Europe sided with Germany, French president Charles de Gaulle held out for SECAM's success. France had watched its computer industry become dominated by American companies and technology, which complicated its efforts to develop an independent nuclear arsenal. If it wanted to retain any international stature, France needed to preserve its electronics industry; De Gaulle used the SECAM patent to assure that outcome. He forced acceptance of the system within the French government and consumer electronics industry, and then negotiated an agreement with the Soviets in 1965 to adopt it, along with their eastern European satellites.

As a result, Europeans had a common 625-line standard, but a series of international meetings to settle on a common system resulted in two color systems compatible only by expensive processing. Germany began its PAL color broadcasts in August 1967, followed by the launch of France's

SECAM programming in October. Over the next twenty years, the rest of the world's countries adopted primarily PAL or SECAM systems, largely based on political and cultural relationships with the early sponsors.

By 1966, then, the structure for broadcast television around the world had been established. That fall, David Sarnoff basked in an industrial tribute to his sixtieth anniversary in electronics and communications. After spending $130 million to make color television part of the American and ultimately international standard of living, he could regard it as a crowning achievement in a 60-year career, and still look forward to "more dramatic progress than any comparable period we have known" (Sarnoff, 1968, 297). Although he left the definition of that progress unstated, Sarnoff was correct. For, by 1971, three companies would announce significant inventions for the future of video technology: RCA's liquid crystal display, AT&T's charge-coupled device, and Intel's microprocessor. These would complement other ongoing innovations in systems stimulated by the scope and scale of the television industry. To these and their causes and consequences we turn in the next chapter.

5

Children of the Revolution, 1947–1987

◆

Television is a system: an interactive, interdependent cluster of people and technologies organized to achieve specific tasks. As individuals and organizations commit time, money, and skills to systems, they stimulate the formation of more systems of new technologies, new industries, and new cultures. When television passed the tipping point of a successful commercial system in the early 1950s, its inventors and entrepreneurs already sought ways to improve its scope or reach. These included new technologies for transmitting, capturing, recording, and displaying live video. As a result, by the last twenty years of the twentieth century, more people could produce and watch more varied video content in more environments than most of their predecessors could have possibly imagined.

Who was responsible for these developments? Unlike the innovation of electronic monochrome and color television, the answer was no longer David Sarnoff and RCA. To be sure, Sarnoff's vision and the company's engineers and scientists continued to contribute to the improvement of consumer electronics and communications. Over the rest of the century, however, domestic and foreign inventors and entrepreneurs from corporations, governments, and garages all joined in developing the next generation of video technologies. Three factors accounted for RCA's diminished role as the twentieth century neared an end.

First, the United States Department of Justice continued its thirty-year pursuit of antitrust action against RCA and other corporate monopolies. In 1958, RCA signed a consent decree that ended its policy of package licensing its 10,000 broadcast-related patents, and gave them freely to all American manufacturers. Besides reducing RCA's incentive to continue research in broadcasting, the decree stimulated more aggressive licensing of its technologies to European and Asian companies. Japanese engineers and entrepreneurs became adept at innovating RCA and other American companies' laboratory demonstrations and exporting them back to the United States as well-designed, inexpensive, imaginative products. Asian corporate investment, Asian government subsidies, American arrogance, low Asian labor costs, and low import duties all helped drive domestic companies and their employees out of the radio, phonograph, television, video recording, video camera, and display industries after 1965.

Second, the Cold War with the Soviet Union affected opportunities for American researchers and corporations. Waging an undeclared "war" boosted government funding on research that might prove useful militarily a generation later; state-of-the-art design made current military technologies as efficient and useful as possible. Military contracts saved corporations the cost of capital investment in equipment and facilities while they guaranteed a net profit of about 6 percent. In a company like RCA, where commercial investments in color television and computers took years to pay off or never did, these contracts provided a steady income, maintained business volume, and maintained the company's technical skills. It also meant managers and technical staff turned their attention away from improving current consumer technologies or innovating new ones. Without the distraction of military markets, Asian and European companies focused on the transfer and improvement of advanced technologies to commercial and consumer markets.

Third, David Sarnoff reached the end of his career and life in the middle of this period without leaving an equally visionary successor in his place. Before giving up control of RCA in 1966 and dying in 1971, Sarnoff insisted that his oldest son succeed him. Robert Sarnoff tried to overcome the burden of his father's legacy by innovating a prerecorded home-video system, taking on IBM's monopoly in industrial computers, and making the company more profitable. It is difficult to accomplish three goals simultaneously and this was no exception. The path to short-term profit appeared to be through diversification in unrelated businesses, offsetting the cyclical economics of the electronics industry. That was true, but it added debt while draining investment in, and attention to, new and core technologies. Investment in computers against a company controlling 90 percent of the market had the same effect. RCA's board of directors fired Robert in 1975 and failed to find a durable successor over the next ten years.

The innovation of broadcast television's descendants overlaps the changes described in the previous and following chapters. Over forty years, entrepreneurs at home and abroad, in corporate and grassroots environments, developed cable and satellite television systems, video cameras, video recorders, and video displays into significant industrial systems beyond their origins in broadcast television. Consumer purchases of their innovations improved the quality and expanded the quantity of video content for growing audiences. Their success affected the monotony of network broadcasting by offering more and specialized outlets for program content. Their commercialization increased flexibility in seeing at a distance through forms of transmission, image capture, and image display.

THE CABLE ALTERNATIVE TO BROADCASTING, 1948–1969

By the end of the twentieth century, some 75 percent of American households paid to receive their television signals through cable connections, rather than tune in for free over the air via antennas. What began as a local effort to provide access to some television channels in remote locations evolved into a global corporate system offering hundreds of choices. For the vast majority of viewers, this freedom—within the monopoly price of a community's cable franchise—was a benefit that trumped its cost and the availability of broadcast channels.

English cities remote from London established local cable stations in 1938. In the United States, the process started when millions of Americans started buying television receivers in the late 1940s. In separate, remote areas of the country, local entrepreneurs developed similar methods for retransmitting distant television signals to homes in their communities. Instead of receiving broadcast television signals or channels through antennas inside or outside the home, a household receives them through a cable that connects to the television's tuner. The other end of the cable runs to a local distribution node called a headend. That collects television channels through broadcast, microwave, or satellite antennas, and processes and combines them for distribution through coaxial or fiber-optic cable that subdivides to each customer. Piping television channels through the cables and adapting them for the receiver's tuner require a variety of engineering techniques to amplify and keep the different signals undistorted by each other, changes in climate, or attenuation between amplifiers.

The first known cable entrepreneur, retailer Peter Walsonavich, bought a television sales franchise in the Appalachian Mountain town of Mahanoy City, Pennsylvania. He then found that he and his customers could watch Philadelphia stations only from the top of one of the surrounding mountains. In 1948 he connected an antenna there to his store and began providing free cable television service to give his customers something to watch. A year later E. L. Parsons of Astoria, Oregon, requested approval from the Federal Communications Commission (FCC) to redistribute broadcasts from Seattle's KING television station by cable. J. E. Belknap & Associates requested the FCC's permission in 1951 to use point-to-point microwave antennas to retransmit signals from distant stations in St. Louis and Memphis to three towns in Missouri and Illinois. By 1960, 640 small businesses provided Community Access Television (CATV) to 650,000 paying subscribers.

With its charge to regulate and promote the public interest in broadcasting to all Americans, the FCC found these grassroots systems puzzling. Nothing in the 1934 Communications Act covered them. At the beginning they provided television service to otherwise isolated communities. As cable entrepreneurs became more ambitious and technologically creative, however, they reduced the audiences for local television stations by offering more channels, distant stations, and better reception because of the local amplification of the broadcast signals. In addition, microwave transmissions made possible exclusive, fee-based special events, like boxing matches or recent movies. Local politicians found them useful for charging franchise fees, and sometime bribes or kickbacks to gain or maintain the local franchise. Successfully pressured by the National Cable Television Association (NCTA), Congress refused to act on the question of regulation. The FCC did not assert its jurisdiction until 1965–1966, when it issued its First and Second Reports and Orders on CATV.

These regulations balanced the interests of broadcast stations with the commission members' ideal of localized community television, which had virtually disappeared from the broadcast market. They assumed that cable was a supplement, not an alternative, to broadcast television. Therefore it should have its own identity as a medium for the exchange of information in the public interest. The first Order required CATV businesses to carry local stations if the business operated within the area covered by the authorized reach of those stations, and limited the transmission of outside programming that duplicated content on local stations. The second Order kept CATV out of the top 100 television markets in order to protect struggling UHF stations in those larger cities. In 1969, following from the monopolistic nature of cable franchises and their growing combination by multiple system owners, the FCC further mandated that cable operators provide public-access channels for local governments, schools, and generate their own local content as well. To further emphasize the local television culture by limiting cable's appeal over local stations, it prevented cable stations from showing movies less than ten years old or sports events broadcast within the previous five years.

WHEN CABLE MEETS SATELLITE, 1945–1985

During the 1970s, the NCTA fought these regulations in the courts. Its lawyers gradually succeeded in identifying CATV with the publishing rather than the broadcast industry, which resulted in fewer restrictions on cable television's content and location under the First Amendment of the Constitution. The other major enabling factor in the diffusion of cable TV as an entertainment utility was the commercialization of satellite communications.

INVENTING SATELLITE COMMUNICATIONS, 1945–1965

In 1945 a British Royal Air Force engineer named Arthur C. Clarke wrote an article predicting the development of geosynchronous, or geostationary, communications satellites. Such satellites orbit the earth 22,300 miles above the equator at the same rate as the earth rotates, keeping them in the same location relative to the earth's surface. This reduces the cost of tracking a satellite from the ground, and of sending and receiving its signals. Building on earlier theories and the electronic and rocketry innovations of World War II, Clarke spelled out the technical and physical requirements as well as a number of unsolved problems. Thirty years later, a combination of government and commercial engineers, entrepreneurs, bureaucrats, and politicians innovated commercial geosynchronous satellites capable of retransmitting video from one source to multiple receiving stations.

Because rockets had military applications and satellites involved international space and signals; because the project was so expensive; and because the United States was immersed in the Cold War with the Soviet Union, the U.S. government funded most of the development. Together with the military, the National Aeronautic and Space Administration (NASA) spent billions of dollars on space technology. The first of many challenges involved constructing reliable rockets that could attain the desired altitudes with a practical payload. After the Soviets orbited Sputnik satellite in 1957,

American politicians, generals, and editors in the mass media fueled investment through NASA in what became the Space Race.

Within that contest lay the question of how to use space for nonmilitary purposes. Several American companies invested in satellite development on the assumption that the U.S. government would continue, in the words of outgoing President Dwight Eisenhower in 1960, to "encourage private industry to apply its resources toward the earliest practicable utilization of space technology for commercial civil communication requirements" (Glover). In particular, AT&T spent $70 million on Telstar, the first active communications satellite. Live television transmission via satellite began with Telstar I in 1962. A thousand managers, engineers, and scientists designed and built ground stations in England, France, and Andover, Maine; solar cells and radiation-resistant transistors; transponders; and traveling wave tubes capable of amplifying signals at gigahertz frequencies. This was the first step in the company's ambitious plan to place fifty or more satellites in low-altitude, non-synchronous orbits 7,000 miles in space, enabling telecommunications coverage of the entire earth. While not as ambitious as AT&T, RCA and its chairman, David Sarnoff, anticipated a significant expansion in personal and broadcast communications with the development of satellites. Both companies transmitted television across the Atlantic, and RCA's Relay II initiated color coverage across the Pacific Ocean, of the Tokyo Olympiad in 1964, to the United States and then Europe.

Under presidents John Kennedy and Lyndon Johnson, however, Congress chartered a private corporation, COMSAT, to lead the country's developments in international space communications. Ironically for Sarnoff, who lobbied for RCA's selection, COMSAT resembled his company as it was founded in 1919: a government-approved monopoly on U.S. involvement in global, point-to-point, wireless communications. One result was that after AT&T finished its experiments with Telstar I and II, it abandoned the development of synchronous satellites to COMSAT and its contractor, Hughes Aircraft.

INNOVATING SATELLITE TELEVISION, 1965–1980

In cooperation with the International Telecommunications Satellite Organization (INTELSAT), COMSAT launched the first commercial geosynchronous satellite, Intelsat, or "Early Bird," in 1965. While COMSAT and INTELSAT expanded the transmission capacity and durability of such satellites over the next six years, the Soviet Union began using its Molniya, or

Lightning, military satellites in special elliptical orbits for its domestic television service, Orbita, in 1967. Five years later, the Canadian Broadcasting Company launched the first domestic geosynchronous satellite, Anik A-1, built by Hughes Aircraft and launched by NASA, to provide color television to all of the second largest country's scattered inhabitants.

The limits on applications and a limited vision for future uses delayed American developments to some degree. In 1965, Sarnoff foresaw the ability "within a decade . . . to broadcast directly into the home from synchronous satellites" (Sarnoff, 1968, 182). He made a valid point: there would be great demand for television through the opportunities offered by satellite technology. Only in 1972, however, did the FCC authorize any individual or organization to operate a satellite, at the same time that it continued to restrict AT&T from using satellites beyond voice and data communications. Western Union, which wanted to diversify its services, received the first approval and launched Westar I in 1974.

RCA's domestic communications division also sought to diversify and its vice president, Howard Hawkins, convinced the company to form Americom to operate RCA's satellite business. Unlike most of the men running the business who thought of point-to-point communications as voice and data transmission, Hawkins realized that the amount of bandwidth needed for video transmission would be extremely profitable. Thus he negotiated a contract with a struggling cable-TV company called Home Box Office (HBO) to distribute its programs from one location to CATV systems across the country.

Satellites offered a relatively inexpensive solution to the problem of national distribution of exclusive cable programming, especially Charles Dolan's concept of pay-TV for premium programs. To increase revenue for his franchise on the island of Manhattan in 1972, Dolan had Gerald M. Levin report on the practicality of microwave networking of cable operators for pay-TV services. Levin's optimistic answer led to HBO's formation; its first program that November was an ice hockey game that attracted 365 subscribers in Wilkes-Barre, Pennsylvania. Three years later, 280,000 subscribers in the northeast United States watched HBO via the expensive relays. With the RCA contract, however, HBO stimulated sales and decreased distribution costs. In September it delivered its first satellite program, the "Thrilla in Manila" boxing match from the Philippines between Muhammad Ali and Joe Frazier. HBO paid for the FCC-mandated 30-foot satellite reception dishes at each operator's headend and sued the FCC for overstepping its regulatory boundaries. This time—unlike with color television standards in 1951 or cable television in the 1960s—the courts agreed with the technological entrepreneurs.

Freed of regulatory barriers and aided by innovations in satellite relays and a steady expansion in the number of channels offered, cable television bloomed in the late 1970s and early 1980s. HBO's subscribers doubled in number in each of the four years after 1975; cable subscribers overall quadrupled from ten to forty million between 1975 and 1985. Each of them had a set-top box that linked the cable to the receiver, and permitted access to one of the many channels paid for through channel 3 or 4 on the television's tuner. Ted Turner helped fill basic services for thousands of cable providers by leasing a satellite channel to extend the reach of his tiny UHF television station, WTBS in Atlanta, across the country. He followed with the Cable News Network (CNN) in 1980, while Brian Lamb started C-Span the year before, providing live coverage of government operations in Washington, DC, Marion (Pat) Robertson stepped up his evangelically oriented UHF station in Virginia Beach, Virginia, to national coverage via satellite in 1977, followed two years later by a group of sports-loving entrepreneurs who underwrote ESPN in Bristol, Connecticut. Larger corporations joined in by establishing USA Network in 1977, a family entertainment channel similar to the three traditional broadcasters.

HOME SATELLITE TELEVISION, 1975–1990

All of these channels and dozens and then hundreds more became available to people willing to pay for basic and premium packages of channels—and the advertising that providers began to insert as well. One consequence of beaming cable channels from satellites whose "footprint" covered entire continents, however, was a homemade revolution in satellite reception. British Broadcasting Corporation technician Steven Birkill built his own dish antenna in India for a satellite transmission test in 1975, fulfilling Sarnoff's prediction. Three years later *Community Antenna Television Journal* editor Robert Cooper did the same at his home in Oklahoma. He wrote an article for *TV Guide* magazine, publicizing the possibilities to millions of television owners.

The research he led on antenna reception led to an FCC decision in 1979 that ended the need for $125,000 licenses for satellite receiving stations, or for 30-foot antenna dishes. Thoughtful amateur and professional engineers exploited the satellite downlinks on the C band of the electromagnetic spectrum with dish antennas 12 feet in diameter or less, if they were willing to accept a degraded signal. These enterprising men—and most were male—throughout the rural United States built and installed "Big Ugly Dishes" in their yards, alternately annoying, attracting, and inspiring their neighbors. Analysts estimated some 1.5 million homeowners installed dishes between 1980 and 1985, with another 300,000 contributing to a thriving market in dish equipment.

That new industry and its customers protested loudly to Congress when the cable companies responded by scrambling their satellite signals in 1986. Only subscribers with a descrambler could receive the companies' channels; this created another running battle and cottage industry in encryption and decryption technologies. The TV-Receive-Only (TVRO) dish owners responded that requiring payment of programs and channels already supported by advertising was greedy, and that in any case a signal sent over the public airwaves was fair game. During the late 1980s the cable industry tried to regulate the new industry of descramblers without government regulation and largely succeeded, although a hard core of users continued to test and share methods to receive satellite programming for free.

Cable and early satellite television stimulated a variety of cultural consequences. They created new industries and jobs, not all of which were applauded. The "cable guy" was one of thousands of technicians with varying aptitudes for the technologies and aesthetics of home installation in 60 million households by 1990. He became a pop cultural icon immortalized in the eponymous 1996 film starring Jim Carrey, and perpetuated in blue-collar culture by comedian Dan Whitney's alter ego, Larry the Cable Guy. The explosion of channel options gave rise to terms like "couch potato," coined in 1979. The phrase "channel surfing" followed in 1986, with people clicking the wireless remote controls that consumer electronics companies had embellished since Zenith Radio Corporation introduced the "Lazy Bones" Flashmatic in 1955.

More significantly, the need to fill the demand for content reflected and spurred the fragmentation and diversification of American culture, as well as the revival and recycling of older movies and broadcast television shows. Reprising Newton Minnow's comment on the "vast wasteland" of network programming a generation before, Bruce Springsteen began singing "57 Channels (and nothin' on)" in 1992: but now there was nothing for everyone's taste. CNN predated the Internet in creating the continuous news cycle, which accelerated the pace of reporting on events and responses by those involved. Satellites increased the flexibility of the broadcast networks as well, most obviously through the use of live remote reporting for local and international news. The need to compete with the virtually unregulated content of cable companies drove broadcasters to provide programs that pushed the bounds of decency as the FCC defined it. The market share of eyeballs for ABC, CBS, and NBC declined, but they still retained the biggest percentages of American households watching TV. Consequently the traditional networks continued to net profit margins approaching 40 percent into the twenty-first century from sponsors seeking the largest audiences.

VIDEO CAMERAS, 1950–1988

Providing broadcast video from remote locations required the miniaturization and portability of television cameras and their associated equipment with little decline in quality. This process also started in the late 1940s and continues to the present. Beyond broadcasting, miniaturization also includes a trend of declining prices while providing a range of resolutions. This has contributed to a culture of surveillance in western societies, in which people increasingly find themselves on camera—sometimes willingly, sometimes involuntarily, and sometimes unconsciously.

INNOVATING MORE VERSATILE CAMERA TUBES, 1947–1980

In 1947, as RCA began to promote the sale of broadcast cameras using the image orthicon tube invented during World War II, one of its inventors, Paul Weimer, sought a simpler, cheaper, camera tube. Up to that point, researchers had worked with photoemissive materials for imager surfaces rather than photoconductive substances like the selenium that first provoked interest in television in the 1870s. At that time selenium proved impractical for moving images because of its sluggish photoelectric reaction. During World War II, however, the American and German militaries sponsored significant research and development in finding and processing photoconductive materials sensitive to infrared light, just below visible light in the electromagnetic spectrum. Drawing on that research, Albert Rose led a group at RCA Laboratories in Princeton, New Jersey, to study photoconductors with a high gain in electronic activity when illuminated with visible light, while Weimer and Stanley Forgue experimented with evaporating thin films of photoconductors on glass substrates coated with a transparent electronic conductor.

Simultaneously with other researchers testing materials for scanning in xerography, they discovered to their surprise that "beautiful glassy red" selenium—that is, amorphous selenium with no structure to its atoms' arrangement—was photoconductive (Weimer, 1976, 743). An extremely thin layer of it on the front of a vacuum tube offered excellent sensitivity to light and surprisingly little lag, or smear, of moving images. As a result, RCA's 1950 vidicon camera system needed no additional multiplier stages to amplify the signal: only a low-velocity electron beam to discharge the electrons gathering on the back side of each cell. After several hundred hours of heated use in the tube the selenium became metallic, however, leading Forgue and Richard Goodrich to develop a durable alternative from antimony trisulfide.

Ironically the images from this sensor smeared in conditions of too much light and suffered from burn-in of the image when left focused on one scene. But it made a portable and versatile camera for a wide range of nonbroadcast purposes. Numerous companies made and sold vidicon cameras for applications like prison and corporate security, satellite surveillance, and Neil Armstrong's first steps on the moon in 1969. Lunar cameras were not cheap; the first one built by Westinghouse Electric Corporation for Apollo 11 cost NASA $453,000.

While vidicons provided the basis for nonbroadcast video for over thirty years, RCA's image orthicons monopolized broadcast applications until 1965, when CBS demonstrated cameras loaned by Royal Philips Electronics of the Netherlands. The dramatically better imagery appeared through plumbicon tubes. Physicists and chemists at Philips Research Laboratories in Eindhoven spent several years looking for a better target for vidicons by methodically selecting and testing photoconductive materials until they arrived at lead oxide. With even more sensitivity and less smear, the compact plumbicon fulfilled Weimer's dream of a simple broadcast camera, although the craft used to build the prototypes in Philips's labs proved difficult to replicate in the factory.

Together with reductions in the size of image sensors, innovations in semiconductors from discrete transistors to integrated circuits (ICs) made miniaturization of high-quality cameras a matter of time and the market available at a profitable price. Combining a vidicon with miniature vacuum tubes in an 8-pound camera and a 53-pound backpack microwave transmitter, RCA Laboratories unveiled in 1951 the possibilities for electronic news-gathering (ENG). NBC reporters used the "walkie-lookie" and the transistorized "creepie peepie" at the presidential conventions in 1952 and 1956, but RCA then abandoned portable, commercial video cameras for twenty years.

Instead the Westel Company in Redwood, California, began selling its portable monochrome TV camera, the Video Cruiser, with associated equipment in 1966 for $10,500. Westel's engineers used ICs to replace bulkier electronic components and reduce the camera's weight to seven pounds. With the networks switching to all-color broadcasting, however, the market for monochrome was declining. Thus it was only unique situations, like capturing live video from outer space, where monochrome made sense. In 1967 RCA's Astro Electronics Division, drawing on NASA's extensive budget, demonstrated a 2-pound monochrome camera for use on the Apollo space missions. Color remained pricey for portable applications; the same year, Philips sold its 23-pound Norelco PCP-70 color camera for over $40,000 to military and educational broadcasters. Others offered different options for ENG broadcast use, but RCA's Broadcast Division in Camden, New Jersey, dominated the field for the last half of the 1970s with the $37,000 TK-76. A camera popular for its portability, durability, and sensible design, networks and stations bought over 2,000 of them for studio as well as portable production. Together with power supply, it weighed 27 pounds and eliminated the use of a backpack.

At the same time, however, Japanese companies began introducing studio cameras in the United States. Hitachi and the NHK Laboratories collaborated on another improved vidicon, the saticon, whose RCA-conceived target comprised selenium, arsenic, and tellurium. By 1980, Hitachi, Sony, Ikegami, and other Japanese companies dominated the American market for broadcast equipment and for home video cameras. That year the cost of vidicons and associated electronics had dropped in price sufficiently for Sony to offer the first video camera intended for the home user.

SOLID-STATE VIDEO CAMERAS, 1965–1985

In tandem with the compression of electronic components through the use of transistors and ICs, several companies pushed the use of semiconductors to replace the camera tube. This was the largest and most expensive component of television cameras; it also used the most power. After the invention of ICs in 1958, RCA, Westinghouse, and other companies drew on military funding to research and develop light-sensitive arrays of semiconductors that scanned themselves, rather than through an electron beam. Since ICs in the early 1960s consisted of 25 to 500 components, fabricating an array of 250,000 sensor elements for broadcast quality imaging was ambitious goal. Nearly twenty-five years and one significant change in technology passed before companies in Japan and the United States could sell commercial cameras using digital chips.

Having spent much of the 1950s trying to make a camera tube that would capture color the same way that the shadow-mask cathode-ray tube provided color for displays, Paul Weimer took a sabbatical at the Sorbonne in

Paris to master solid-state physics. AT&T's Bell Laboratories started full-scale study of that field when it announced the transistor effect in 1948. Instead of manipulating electrons in a vacuum, solid-state research involves moving electrons, and the "holes" they leave when moved, in the atomic structure of materials. Weimer realized that semiconductors offered applications for his "first love," imaging technologies, and returned to RCA's Princeton labs in 1960 to apply his latest education to his experience with camera sensors (Weimer, 1975, 20).

His small team reduced thin-film transistors (TFTs) to practice in 1961. Believing that silicon lacked the photoelectric properties suitable for a camera sensor, Weimer drew on thin-film electrodes in vidicons and a series of Air Force contracts to build a solid-state camera using cadmium-selenide TFTs. Westinghouse's researchers proved Weimer wrong by demonstrating a camera with an array of 50×50 silicon phototransistors in 1965. His group scaled up to 180×180 elements, or pixels, using the world's largest IC in 1967, and transmitted images with a 256×256 wireless camera the following year.

Both the silicon and TFT arrays used matrix addressing of the X rows and Y columns to scan the light information at each transistor. In 1969 and 1970 F. L. J. Sangster and K. Teer of Philips and Bell Labs' Willard Boyle and George E. Smith reported solid-state scanning by charge transfer or, as Sangster's method became known, by "bucket brigade." Here the array was scanned line by line, migrating stored photoelectric charges into registers and the camera's output circuit. The result was a low-power, low-noise, high-sensitivity image from a small, durable chip.

Boyle and Smith conceived their charge-coupled device (CCD) while imagining an alternative form of computer memory, but Smith had worked before on AT&T's Picturephone project. Over forty years after the telephone company's first experiments and demonstrations, its staff still pursued the prospect of people seeing each other on television while they talked by phone. The prospect of a solid-state camera chip gave Boyle and Smith the funding to build and demonstrate monochrome and color CCD cameras in 1971 and 1972. When AT&T abandoned Picturephone and the CCD in the latter year, other researchers and managers at Sony, Fairchild Semiconductor, and RCA found the technical possibilities exciting enough to pursue further development.

Advocates at the American companies, however, found little corporate interest for development of the technology in a video camera. Fairchild's Gil Amelio foresaw the mass production of CCDs for home cameras, as did Weimer. Both received funding for specialized sensors for the U.S. Navy

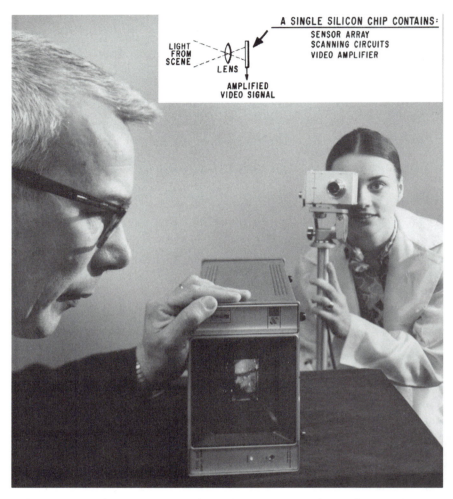

A SINGLE SILICON CHIP CONTAINS:
SENSOR ARRAY
SCANNING CIRCUITS
VIDEO AMPLIFIER

LIGHT FROM SCENE
LENS
AMPLIFIED VIDEO SIGNAL

Technician Lynn Roberts captures Winthrop Pike at RCA's David Sarnoff Research Center with an early CCD camera in March 1972, two years after AT&T announced the technology. The use of photosensitive solid-state elements eliminated the bulk and delicacy of vacuum tubes and their glass envelopes. David Sarnoff Library

and Air Force. When the contracts ran out, Amelio teamed with Ricoh to make CCD line scanners for copiers and fax machines before the recession in 1974 cut that line of support.

At RCA, Weimer switched from unstable TFTs to Sangster's bucket brigade. Together with Frank Shallcross and Winthrop Pike he built RCA's first charge-transfer camera, a 32 × 44 array that showed recognizable faces. Weimer finally turned to CCDs, where Walter Kosonocky had been leading an RCA team that pushed the dimensions of solid-state imagers to 512 × 320 pixels by 1974. In 1975 Weimer forecast that "if we have these kinds of cheap cameras and if we have a cheap method of storing the videos, then you see we are running with Eastman Kodak...a product, which would really outsell any of the present kind of cameras" (Weimer, 1975, 22).

For all of the advances, however, each of the cameras remained demonstrations of principle. For each CCD chip that worked, researchers rejected thousands of flawed imagers. Eventually RCA's Broadcast Division and NBC introduced the first commercial broadcast camera to use CCDs in 1984. During the World Series that fall, the CCD-1 camera captured the stitches on the baseball as pitchers threw to batters. It worked well in low-light conditions and smeared no images, and served as an ironic coda to over 50 years of leadership in broadcast technologies. The division booked orders for 50 of the cameras at $40,000 apiece and won an Emmy award for technical achievement just before RCA closed its Broadcast Division a year later.

SONY AND THE CCD VIDEO CAMERA FOR CONSUMERS, 1973–1988

Although RCA could not afford or chose not to invest in the technology because of its struggles with leadership and other innovations, Sony Corporation spent $200 million over eleven years to commercialize the CCD camera. In 1973, Kazuo Iwama took over Sony's research center in Yokohama before rising to presidency of the company. One of the first technologies shown to him was an 8-line CCD imager that displayed the letter "S." To help revive the center's semiconductor program, he charged a group of 40 researchers with turning that first CCD into the first mass-produced video camera within five years. "Our main rival is Kodak," Iwama told the staff. "We will compete with film, and it will be a new business for us" (Johnstone, 1999, 202).

Over the rest of the 1970s, Sony's technical staff under Makoto Kikuchi struggled to design, install, and master processing equipment and the clean rooms that contained them. Impurities in the silicon chips' lattice structures generated excess electrons, fouling the signal. Worse yet, a speck of dust covering even part of an element affects the brightness in display, to which the human eye is more sensitive than color. After three years of failures, Iwama faced down challenges from Sony's board of directors to keep the project going. By the deadline, in 1978, Sony demonstrated its first practical

CCD camera at a press conference and moved the project to its factory in Atsugi outside Tokyo.

The first application, for 52 of the CCDs, was for cockpit-based cameras on All Nippon Airways so that passengers could watch take-offs and landings. Because the yields from each wafer of chips were still low, each chip alone cost over $1,600. Iwama died in 1982; Sony began selling its CCD camcorder in January 1985. After a year alone in a home video camera market dominated by vidicon tubes, Sony was joined by JVC and then by Sharp Electronics. By 1994, Sony manufactured ten million CCDs a year and controlled over 40 percent of the camcorder market. Camcorders, which combined a camera and videotape recorder in one device, accounted for 85 percent of all CCDs produced, and sales forces and consumers expanded markets into video conferencing and digital cameras that can store both still and moving images.

COMMERCIAL VIDEO RECORDING, 1951–1980

At the same time that Weimer and his groups were reducing the vidicon tube to practice in a portable camera in 1950, CBS's chief engineer Howard Chinn suggested an alternative to the networks' practice of recording programs on film. He asked, "Why not store the video signal on magnetic tape? . . . Such a scheme would use up a lot of tape, but it might well be worth it" (Abramson, 2003, 37). Chinn touched only the surface of what was possible with what came to be known as videotaping. Recording the video signal directly would not only free it of the noise that transferring images through lenses and chemicals entailed; based on the systems in place for radio for the previous four years, magnetic recording would enable electronic editing, replays, ENG, and far more flexibility in what television programmers could show, and when they could show it.

If one can record or transmit content, then one should be able to transmit or record it, respectively, as well. John Logie Baird recorded the first television signals in 1927, in snippets on 78-rpm gramophone discs that he had no way of playing back effectively. Higher-resolution images from television waited until the U.S. and German armed forces recorded CRT displays on 35mm film to track the targeting of guided missiles during World War II. In 1945 DuMont initiated civilian recording of television off the face of the CRT, or kinescope, with 16mm film cameras. Flagship stations shipped or mailed these "kinescopes" to television stations not networked by AT&T's coaxial cable system or that ran a different schedule. By 1951, New York television studios processed over 46 million feet of film every month and shipped thousands of hours of programs across the country, enabling Californians to watch Christmas programs in January. Recording shows, for repeat viewing in synchrony with time zones by coaxial cable or microwave relay, would unify American programming and lower production costs significantly.

VIDEOTAPE AT RCA, 1951–1956

CBS was not in a position to act on its chief engineer's wish, but RCA was. In September 1951, when RCA named its laboratories in Princeton after him, David Sarnoff made three wishes for his fiftieth anniversary in electronics five years hence. One of them was to record color video on "an inexpensive tape, just as music and speech are now recorded. . . . Such recorded television pictures could be reproduced in the home, or theater, or elsewhere at any time" (Sarnoff, 1968, 250). Because the Labs' staff had led the innovation of electronic color television and worked effectively with the manufacturing divisions of the company, it was not unreasonable to expect that it would work similarly on these projects.

The task fell however to Harry Olson, director of the Acoustics Laboratory, whose staff worked on magnetic sound recording. One of the world's leaders in acoustic engineering, Olson had not been involved in the color television effort and the emphasis on sight over sound as the focus of RCA's business overshadowed his status within the company. At the same time, the RCA Victor Division's Advanced Development Group in Camden, New Jersey, retained a number of experts in audio and electromechanical engineering who innovated a variety of audio technologies without the Acoustic Lab's assistance. Olson's pride prevented him from consulting his video colleagues in Princeton or Advanced Development on the challenges involved in recording television, and he ignored the patents and technical literature on the subject, which ran back to 1928.

The challenges of videotaping differed significantly from those in audiotape. For much of the second half of the twentieth century, the best professional sound recording involved converting an electronic audio signal into a changing electromagnetic field. Emitted from a fixed head, the field modulated or changed the orientation of magnetic particles coating a tape that ran between two reels. When played back, the magnetized tape changed the electromagnetic field on the same or a different tape head, which sent the signal to an amplifier or another output.

The biggest difference between audio and video signals was the quantity of information to be recorded. A video signal used about 250 times more bandwidth of frequencies than audio, or two orders of magnitude. In the early 1950s, magnetic recorders required about an inch of tape per second to record 1 kHz of sound. To record high fidelity sound, up to 15 kHz on tape a quarter-inch wide, standard broadcast tape recorders passed 15 inches of tape per second past the magnetic recording head. Project this out to the 4 MHz required for video, and the system requires a tape speed of 227 miles per hour.

In addition, engineers needed to maintain a constant speed past the tape head in order to avoid wow and flutter on playback. Plastic tape stretches and slips during recording and playback, which distorts color video signals to an even greater degree than audio. Finally, magnetic tape is not a linear medium: that is, a constant increase in frequency input does not result in an equal increase in frequency output. The megahertz frequencies involved in video prevented the use of audio techniques to maintain consistent linearity of grayscale and hue reproduction as a result.

How, then, do you pack the additional information onto a tape? The obvious way is to run the tape faster past the recording head, but that uses a lot of tape. Another method is to use multiple recording heads, but that complicates reassembly of the signal. Nonetheless, Olson believed that a high-speed, multihead approach was both practical for studio use and cheap enough to commercialize, compared to more complex solutions. To record on 2-inch-wide tape, his group split color video signals onto five tracks using five magnetic heads, which helped reduce the tape speed to 20 and then 13 miles an hour.

By December 1953, Olson's team recorded four minutes of NBC color programming on a reel of tape 17 inches in diameter; a technician equipped with leather work gloves and goggles helped stop the reels and rewind the tape for playback. Two years later, the reel was 20 inches in diameter but held 15 minutes of content. Some of it showed very good imagery, but there was little practical about the technology. Despite the criticism of other members of the research center staff, RCA's Broadcast Division engineers set about trying to make Olson's video recorder ready for prime time.

VIDEOTAPING AT AMPEX, 1951–1956

RCA never escaped the direction in which Olson led the company. Meanwhile, six young engineers at Ampex Corporation in Redwood City, California, debuted a commercial videotape recorder at the National Association of Broadcasters convention in the spring of 1956. That it recorded only monochrome did not affect sales because only NBC offered color for special broadcasts. As a result Ampex broke RCA's monopoly in broadcast technologies, and stimulated the innovation of video recording for all sorts of uses.

This work began in conjunction with a number of other small engineering organizations. One of the leading experts in magnetic recording, Marvin Camras of the Armour Research Foundation in Chicago, began to consider video recording in 1949. When he concluded theoretically that high-speed tape was impractical, Camras built a recorder where multiple heads rotated within a drum at high speed as the tape passed by. Camras did not pursue this technology, but Walter Selsted of Ampex, which licensed Camras's patents, visited and brought the concept back to California in the fall of 1951. There, Ampex founder Alexander Poniatoff agreed to invest $15,000 in a machine using a 2-inch tape moving at 15 inches per second past the heads.

About the same time, in November 1951, John Mullin demonstrated the first magnetic video recordings in nearby Los Angeles. He had pioneered professional audio tape recording in 1946 and now used the same approach Olson and five other organizations tried in the early 1950s, with a fixed head and high-speed tape. *The New York Times* reported that images recorded from a television broadcast onto standard audiotape were "blurred and indistinct." Nonetheless Mullin insisted that his group was "far enough along to straighten out the snarls" and could offer recording at one-tenth the cost of film (Abramson, 2003, 50).

Selsted hired Charles Ginsburg to reduce Camras's approach to practice. An early demonstration in March 1953 led Poniatoff to ask, "Which is the cowboy and which is the horse?" (Abramson, 2003, 52) Ampex stopped the project until after RCA's first demonstration in December, which helped convince management that it was a technology worth pursuing. Ginsburg's group made little progress until Selsted and Ross Snyder reviewed the available literature on magnetic recording in mid-1954. They concluded that frequency, rather than amplitude, modulation of the signal offered huge advantages in linearity if they could overcome the technical complexity and costs involved. Selsted also suggested recording the signals transversely, at a right angle to the movement of the tape, on four heads located on the rim of a rotating head wheel. This quadruplex approach gave a tape-to-head speed equivalent to tape moving 80 miles an hour while the tape moved roughly 100 times more slowly.

Over the winter of 1954–1955 Ginsburg and his group resolved the challenges in designing FM circuitry and maintaining the tape's contact with the head wheel. By March 1955 their demonstration of a very clean monochrome recording and playback to Ampex's board of directors gave them the support needed to turn their recorder into a product. That led to an engineering demonstration less than a year later inside Ampex. After the unexpectedly large crowd watched the team record a bit of a broadcast and play it back, Ginsberg recalled 30 people shaking "the building with hand-clapping and shouting" and two of the six engineers who had argued the most slapping "each other on the back with tears streaming down their faces" (Inglis, 1990, 322).

Poniatoff and his managers moved quickly, inviting executives from CBS, ABC, the Canadian Broadcasting Corporation, and the British Broadcasting Corporation to another demonstration. CBS immediately offered support in return for a preview for TV stations affiliated with network the day before the National Association of Broadcasters convention in Chicago in April 1956. The word of mouth from the preview resulted in lines of eager engineers in the hallways of the Hilton hotel, and orders for 82 of the $45,000 machines. CBS made the first broadcast from videotape on November 30, 1956, of the 15-minute program, *Doug Edwards and the News*. Engineers taped it in New York, relayed it via cable and microwave to CBS Television City in Hollywood, where other staff recorded and replayed it three hours later for West Coast viewers.

INNOVATING VIDEOTAPE RECORDERS, 1957–1980

When Ampex began delivery beyond prototypes in 1957, the price seemed to be too low. ABC and CBS saved nearly $10,000 a week in costs associated with the kinescope film system. Meanwhile, RCA's George Brown, who watched a preview just after the CBS affiliates, organized a team to adapt Ampex's system to color recording. Once RCA received its first Ampex recorder, 25 engineers took it apart and built a color model. Their demonstration to Ampex in 1957 led to a cross-licensing agreement for recording and color techniques, plus $200,000 to Ampex. That company's monopoly ended in 1959, and RCA introduced the first transistorized video recorder in 1961. Ampex's engineers responded with a recorder that improved the quality of color video by a factor of 10, four years later. By extending the maximum frequency range of the tape recording on the same 2-inch tape, they improved the ratio between signal and noise, and dispersed the components of the signal so that they no longer interfered with one another.

While Ampex, RCA, and other companies continued to improve quadruplex technology, its inherent flaws led their, and aspiring Japanese, engineers to explore alternative methods of recording. Chief among these was helical scanning, in which the tape wraps at an angle around a rotating drum containing a recording head. By recording a complete field of video in one pass of the head on the tape, it promised to reduce the precision required for quadruplex machines. It also enabled more special effects like still frames and slow motion.

RCA's Earl Masterson applied for the first helical patent in 1950, while working on a military project in RCA's Camden, New Jersey, plant; the armed forces funded numerous magnetic recording projects for rocket telemetry and jet aircraft and weapons tests. While Olson ignored its commercial applicability, Alex Maxey at Ampex built the first working helical recorder in 1956. Inspired by RCA's early demonstration of color video recording, Norikazu Sawazaki of Tokyo Shibaura Electric Company spent six years in research and development of a practical helical alternative before leading the first public demonstration in 1959. Two years later Japan Victor Company (JVC) led Toshiba, Sony, Ampex, and Loewe in Germany in introducing the first commercial helical recorder. The prices were usually less than half that for quadruplex machines, in exchange for reduced quality that limited them to nonbroadcast applications.

The capital invested in quad players, the libraries of taped shows, the lack of a helical standard, and tracking problems in helical recorders held up their diffusion. Nonetheless, many companies invested in their development for closed-circuit educational television systems and electronic news-gathering. Through incremental improvements and ICs for digital control, by the end of the 1960s engineers began solving the problems of timing and tracking during recording and playback. ICs also reduced the bulk of recorders, and magnetic tapes using finer, more sensitive metal particles—also demanded by the computer industry for storing data—reduced the dimensions of the tape reels. By the end of the 1970s, television engineers agreed on a broadcast videotape standard developed by Ampex and Sony, and helical recording technology in dozens of formats swept the industry.

HOME VIDEO RECORDING, 1955-1975

The factors of miniaturization and consistency in manufacture that accounted for magnetic tape's success at professional levels carried over to its triumph over rival formats for recorded video in the consumer. In 1956, Olson also showed Sarnoff and the press a color videotape player for the home. This was no more practical than his broadcast tape system, but

RCA announced the domestic application of Harry Olson's color videotape technology as Chairman David Sarnoff's "Hear-See" fiftieth anniversary gift in October 1956. Ampex's engineers beat RCA to the broadcast-station market with their monochrome magnetic-tape system, but the early color demonstrations and patent licenses helped stimulate Japanese engineers to develop practical home systems beginning in the 1960s. David Sarnoff Library

Sarnoff understood the goal was a device costing several hundred dollars that would serve the same role that phonographs or audio tape systems provided consumers of music.

The technologies to make such a player viable in the home would not exist, however, for another twenty years. In the meantime corporate entrepreneurs and inventors put forward other formats using a variety of technologies. No one knew which one would provide acceptable resolution, playing time, convenience, and cost for a mass market. The option of home recording by magnetic tape only confused matters further. After all, despite the number of radio stations broadcasting music available for home taping, audiotapes in the 1950s remained a niche market compared to phonograph records. Neither RCA's nor Philips's tape cassettes had enjoyed great success by the mid-1960s, raising the prospect that consumers would be satisfied with prerecorded movies and instructional films.

As the networks and the public embraced color television, then, RCA's staff at the David Sarnoff Research Center in Princeton began working on home video systems using lasers and plastic tape embossed with holograms. Lasers, which had been first demonstrated in 1960, appealed to young research engineers with little regard for the cost in a mass market or understanding of production engineering. In Europe, Decca's researchers developed and demonstrated a laser-read videodisc. At CBS, Peter Goldmark persuaded Paley to invest in electron-beam recording on film, an equally impractical technology for mass production.

Without a strong relationship with the broadcast or consumer electronics divisions, RCA's research center explored numerous approaches to home video. Only one of these was magnetic tape, which the Princeton staff abandoned in 1974. While RCA pushed ahead through the 1970s under multiple chief executives toward a $400 videodisc system similar to a phonograph, however, Sony under Akio Morita pressed forward from its advances in broadcast magnetic-tape recording.

SONY AND HOME VIDEO CASSETTE RECORDERS, 1965–1975

Morita and his associates pioneered the development of transistorized consumer electronics that helped Japanese companies wipe out American competition in radios, audiotape players, and monochrome televisions between the late 1950s and early 1970s. He was equally determined to establish Sony as the leader in home videotape players: "People do not have to read a book when it's delivered. Why should they have to see a TV program when it's delivered?" (Lardner, 1987, 68)

Sony's experience in solid-state design and manufacturing attracted Ampex into an agreement in 1960. It gave Sony the right to build Ampex-based videotape recorders for the nonbroadcast market in exchange for Sony's transistor circuits. While Ampex continued to fight off RCA in the market for broadcast recording, Sony proceeded to work its way down successive industrial and educational markets to the home. Morita reinvested the profits in the company's engineers and scientists in materials, electronic circuits, motors, and design to make this possible. In 1965, Sony introduced the CV 2000 in two models for $1,000 and $1,250. By recording and doubling the lines scanned in one of the two interlaced fields that comprised a frame of television imagery, Sony's expert on tape recorders, Nobutoshi Kihara, obtained an hour of monochrome video on a half-inch tape reel seven inches in diameter. Sony included a 6-pound camera for home recording, and a timer "to record a television program while the owner is away from home" (Lardner, 1987, 68).

The popular reluctance to thread the tape to the take-up reel drove the company to develop a cassette format. RCA and Cartrivision, among others, had innovated cartridges that eliminated the need to perform this delicate task, but these were either for broadcast applications or part of a complex consumer system. By 1969 Kihara had designed a book-size cassette, with a tape three-quarters of an inch wide. The extra quarter inch and development of finer metal-oxide particles enabled its color capability. A year later Sony freely cross-licensed its videocassette recorder (VCR) patents with Matsushita and JVC, who modified Sony's U-Matic design and adopted it as the first standard format for the home.

When Sony put the U-Matic on sale for $1,300, or over double what it planned, few American consumers struggling in a weak economy bought the new device. Combined with ever-cheaper video cameras, however, the 60-pound machines and their 30-minute, $30 cassettes instead became hugely popular for education, pornography, and mobile TV crews. CBS initiated regular ENG when it combined the recorder with a portable Ikegami camera to cover President Richard Nixon's trip to Moscow in June 1974.

Meanwhile Kihara designed the next generation, pushed by Sony founder Masaru Ibuka's demand that the cassette be shrunk from the size of a hardback book to a paperback. Kihara's breakthrough came in eliminating the blank guard bands on the tape between each stripe of recording through a technique invented in Japan called azimuth recording. This mounts two recording heads off perpendicular to the path of the tape. To prevent color interference, Kihara changed the phase of the color signals, eliminating the blank bands. His group also shrank the recording heads, tape width and thickness, and circuitry; all of these contributed to a 40-pound machine and a paperback-sized cassette offering an hour of recording time.

Sony offered its Betamax to Matsushita and JVC in 1974, with *beta* referring to a brush stroke in Japanese calligraphy thick enough to coat the paper. To the surprise of Morita and the other executives, however, they balked. Matsushita's market surveys showed a popular desire for two hours of recording for sports programs and televised movies. Sony's surveys showed a demand for a cheaper cassette; longer playing time could come later. Morita also rejected the opposition because Sony had already set up its production lines, and because of his confidence in Sony's record of technological and commercial successes, which were based on his instinctive understanding of the American consumer. The Japan and American response to the $1,300 Beta VCRs and their 15-dollar cassettes in 1975–1976 seemed to prove him right. Morita coined the term "time-shifting" and the advertising campaign drove home this selling point. A year later, half the adults in the U.S. knew about Betamax, appreciated the personal flexibility it enabled in their leisure time, and bought cassettes and players in such quantities that Sony could not keep up with the demand.

THE CONSEQUENCES OF HOME VIDEOCASSETTES AND RECORDERS, 1978–1982

Sony's decision had several consequences. One result was that Hollywood studios, led by Universal and Walt Disney Productions, quickly sued Sony for enabling copyright infringement on broadcast programs. The suit surprised Morita and absorbed millions of dollars and thousands of hours of Sony staff's time. If Sony lost, it would be liable for hundreds of millions of dollars in damages, and the entire VCR industry would have to convert to prerecorded tapes. The initial decision in 1979 favored Sony; a court of appeals unanimously reversed it; and in 1984 the U.S. Supreme Court ruled 5–4 that an initial recording at home from broadcasts constituted a fair use of copyrighted materials. Meanwhile, Americans bought nearly eight million VCRs and 160 million blank cassettes that year, mostly unaware that they were one vote from engaging in mass civil disobedience of copyright.

A second result was the development of grassroots video cultures. One of these featured enterprising individuals, gangs, and organizations that recorded films in theatres, copied hundreds or thousands of videocassettes, and sold them on street corners, local stores, market stalls, and on blankets across the world from Philadelphia to Moscow to Mexico City. As this phenomenon realized the worst fears of Hollywood about losing control of its intellectual property, studios became very reluctant to cooperate with the consumer electronics industry in licensing prerecorded videos. Andre Blay, an audiovisual equipment distributor in Farmington Hills, Michigan, was the first person to license movie videos for sale, and Twentieth Century Fox's Steven Roberts was the only person in Hollywood willing to adjust to the demand for home video: "I never wanted to see our company let a new technology come along and bury its head in the sand" (Lardner, 1987, 172). Blay sold over 250,000 tapes by the end of 1978 at $50 apiece, at which time Fox bought him out for $7,200,000 to run the business itself.

Thousands of Americans dove into the business of renting videocassettes. Frank Atkinson, a failed Hollywood actor, had been renting out film projectors and old films to hotels. When he read about Fox's licensing

Mechanical Engineer Hiroyuki Umeda, team leader Yuma Shiraishi, and electrical engineer Yoshihiko Ohta sit behind the 1976 prototype of what they and other JVC engineers turned into the first commercial VHS machine, the HR-3300, the forebear of over 900 million VHS videocassette recorders around the world. David Sarnoff Library, courtesy JVC

deal with Blay, Atkinson borrowed $10,000 to buy Beta and VHS (Video Home System) copies of 50 Fox movies, placed an ad in the *Los Angeles Times*, and was overwhelmed by the response. Working his way around legal concerns, he began franchising his rental business, which other entrepreneurs across the country emulated. Given the challenges in running a small business, it is unclear how many actually made a profit renting and selling videos, but between 1980 and 1982 Walt Disney Pictures, Fox, and Warner Brothers tried to extract more of the proceeds for themselves. They insisted that the owners of Video Shacks, Huts, Castles, and Corners rent tapes and report their earnings rather than buy videos of their films, just as movie theatres did. Led by Atkinson and store owner Rocco LaCapria in Bay Ridge, Brooklyn, the small business owners organized successful boycotts that retained their independence, forcing Hollywood into what became a successful business of direct sales to retailers as well as video stores.

JVC, MATSUSHITA, RCA, AND THE INNOVATION OF THE VHS SYSTEM, 1976-1988

A third result was that Matsushita, which was three times the size of Sony, pressed forward with JVC's 2-hour VHS system, and cleared the field of competitors. JVC's staff had resented Sony's Betamax presentation as "an ultimatum. There was no room for negotiation, no room for exchange of ideas." When JVC and Matsushita showed Sony the VHS player in April 1976, the roles reversed. Since 1971 the head of JVC's video products division, Shizuo Takano, had led a VCR project where he and two engineers mapped out nineteen design goals for a home system. Five years later JVC had a smaller player, larger cassette, marginally less video quality, and 2 hours of recording time. JVC and Matsushita had also lined up other Japanese companies as licensees. When Sony rendered moot the issue of playing time with a 2-hour Betamax, the argument turned to the loading technology for the cassette tape. In the minds of Sony's staff, JVC engineers drew on what they knew from U-Matic patents and the Betamax demonstrations. Morita's conclusion, "It's a copy," further enraged the former partners, who began selling VHS players in the United States in the middle of 1977 (Lardner, 1987, 146).

Despite Sony's success, two-year head start, and licensing of Betamax to Zenith Corporation, JVC and Matsushita's marketers were correct about the popularity of more playing time. In this approach, they enjoyed the support of RCA, whose executives were willing to sell a VCR as long as they did not have to underwrite the development or production. RCA's Consumer Electronics Division had already discovered that its engineers lacked the skills or resources to make a mass-produced VCR player to the tolerances and prices required. The Princeton labs were of no help, in part because its staff was still working on a videodisc and player. During discussions with Matsushita in 1976, however, Chief Executive Edgar Griffiths and Vice President Roy Pollack had one question: could Matsushita's engineers fit three hours of playing time in the same VHS cassette? For the American men who could afford the initial product, recording weekend football games was a vital selling point.

After some discussion among themselves, Matsushita's executives went one better and offered four hours. Worse for the design and production engineers back in Japan, they had to accomplish this in six months. In return, RCA agreed to buy 55,000 VCRs for holiday sales in 1977, and 500,000 to one million players over the next three years. Matsushita's engineering and factory staffs rose to the challenge, and when Sony brought out its 2-hour

Beta system that year, Matsushita followed shortly with its 4-hour VCR. RCA sold them at a loss for $1,000, in part to tie VCRs to purchases of color televisions, on which it enjoyed a healthy profit margin.

Sony was not suffering. It enjoyed the support of the world's largest retailer, Sears, for distribution, and several American companies for marketing; its sales climbed from 400,000 to 900,000. But by the end of March 1978, Americans already bought more VHS players than Betamaxes and RCA alone commanded more of the market than Sony. In 1985 Americans bought over eight million VHS machines made in Japan, mostly by Matsushita, which manufactured VCRs for seven brands in the U.S. Three years later, Morita swallowed his pride and Sony began selling VHS players as well. By 2006, when JVC received an IEEE Milestone for the technology, people around the world had bought over 900 million of the machines.

THE DEATH OF AMERICAN AND EUROPEAN ALTERNATIVES IN HOME VIDEO, 1978–1985

The other competitive casualties were RCA's videodisc and Philips's laser disc. The premise for the staff of the David Sarnoff Research Center (DSRC) seemed promising in the 1960s. The cost of stamping audio records from gobs of plastic was far cheaper than the fabrication of magnetic tape and the recording of albums on each tape. A stylus would be far cheaper than a laser. RCA was heir to the original disc record company and made and marketed records through its record division. Why not design, make, license, and market a video record and player? The DSRC's experts could apply their skills in materials research, mechanical and electronic engineering, and video processing to fit an hour of video on each side of a 12-inch disc, and a player and stylus for it. Beyond the price, targeted at $400 for a player and $20 for a record, RCA would reassert its dominance in consumer electronics with patent licenses and stimulate sales in its color television business.

What appealed at the start became less so as rivals innovated magnetic tape and laser discs. At the same time that the DSRC committed to its project, Philips started work on a laser disc system. It allied with the entertainment conglomerate MCA and demonstrated a version to RCA and other companies in 1972, which refused to join in. Because ICs for digital video processing were not yet available, the highly focused laser beam recorded and read an analog signal, which provided excellent imagery when everything worked. Just before the holidays in 1978, Philips's American subsidiary Magnavox introduced its $700 Magnavision system and $16 discs to great acclaim. Early adopting videophiles bid up prices on the black market before flaws in production became apparent. Magnavox offered only 5,000 of the players in 1979, while consumers flocked to VCRs. Although the Japanese company Pioneer joined in to correct the production problems and revived sales in the late 1980s, the laser videodisc never attracted more than 10 percent of American customers for home video.

By 1978, RCA was stuck with the pride of its laboratories, a technological marvel that performed as promised. A metalized stylus point too small to see with a microscope transmitted the changing electrical capacitance of the conducting disc's spiral groove, 10,000 of which fit in an inch. The video quality approximated that of VHS tapes. Despite the fact that the Consumer Electronics Division had already committed to the VHS cassette, Chairman Robert Frederick agreed with the Labs and patent licensing executives. He committed the company to its videodisc shortly before his conflicts with the board of directors complicated its commercialization. Over two years passed before RCA, CBS, and several Japanese companies started manufacturing and marketing the system.

Like Sony and its Betamax, RCA sold a reasonable number of videodisc players, 165,000 in 1981, with other companies selling another 135,000. But the numbers dropped in 1982 and the price advantage evaporated as millions of people rented three-dollar cassettes for their videotape players. After dropping the price of players to $200 to exploit surprisingly high sales of the discs, RCA's new chief executive, Thornton Bradshaw, announced the end of the innovation in April 1984. RCA lost over half a billion dollars on the effort, shaped as all technologies are by the experiences, skills, and interests of its innovators as well as the alternatives available to consumers.

DISPLAYING VIDEO

All of these innovations in television and video technology required a display. From the end of World War II to the end of the Cold War, entrepreneurs, engineers, and scientists at many companies responded to consumer desires for cheaper, compact receivers with larger, flatter displays. Initially these efforts took to improving the traditional CRT, both monochrome and color, but in the 1960s scientists and engineers began demonstrating alternatives that could display video in a flat display. Realizing that goal took a generation and made an essential contribution to the video culture of the early twenty-first century.

MONOCHROME DEVELOPMENTS

CRTs offer intrinsic challenges to the goals listed above. The larger the glass envelope enclosing the vacuum, the greater the atmospheric pressure on the surface. The divergence in shape of a CRT from the ideal resistance of a sphere means that engineers have to calculate the uneven stresses that result from over 3,000 pounds of pressure per square inch. In addition, the bigger the picture, the greater the challenge to depositing phosphors and ensuring efficient deflection of the electron beam to provide even brightness across the screen. RCA's Advanced Development Group innovated one alternative in the late 1940s by using a small CRT, mirrors, and a Fresnel lens to project an image up to six times larger that on the picture tube.

The bulky home projection receivers survived until companies began manufacturing 16-inch and larger round displays in 1948, including DuMont's 30-inch diameter CRT for its Royal Sovereign set. Because content was transmitted in the 4:3 ratio of 35mm film, forcing the trimming of image corners on round CRTs, marketers boasted in advertisements about viewing area in square inches rather than diagonal dimensions. Zenith made a virtue of round displays on the grounds that they replicated the shape of the human eye. The Hytron Radio and Electronics Company in Salem, Massachusetts, led the innovation of the postwar rectangular CRT in 1949. Westinghouse's engineers boosted the deflection of the CRT's electron beam from 70 to 90 degrees in 1953, which other companies followed with 110 degrees in 1957. This helped increase the width and reduce the depth of the mostly wooden cabinets and home-entertainment consoles occupying the center of many American living rooms.

As more than half of American families acquired televisions in the 1950s, engineers and marketers also innovated compact, portable receivers. The basic challenge was the dissipation of heat from the vacuum tubes that amplified and processed the received video and audio signals. RCA engineers showed off the first transistorized portable television in Princeton in 1952; the prototype transistors solved the thermal issues but the set could receive only one channel no more than 15 miles from the antenna on the Empire State Building. Instead of commercializing this, RCA introduced its line of "Personal" vacuum-tube portables four years later. Despite a vented metal case the size of a shoebox, they overheated easily. Sony and Philco sold the first transistorized receivers in 1960 and by the 1970s Sony led

Gerald Herzog of RCA's David Sarnoff Research Center checks out the world's first solid-state television in November 1952. The battery-powered set and its 5-inch CRT used 37 high-frequency transistors, weighed 27 pounds, and received one channel 15 miles from the transmitting antenna. David Sarnoff Library

other Japanese companies in selling handheld sets using CRTs as small as an inch in diagonal width.

COLOR TUBE INNOVATIONS

RCA's shadow–mask CRT had been a breakthrough as an electronic color display, and it served two generations of TV watchers. Nonetheless, its drawbacks led RCA and other companies to seek alternatives as well as improvements to make it cheaper, brighter, and flatter. The first color CRTs required alignment of a shadow mask and phosphor–coated glass plane inside a tube comprised of a glass faceplate, metal shell, and glass neck with its electron–gun assembly. The phosphors offered very pure colors but little brightness, a condition aggravated by the mask that blocked 85 percent of the electron beams' energy from the phosphors. Because of color television's commercial failure into the early 1960s, the opportunity remained to avoid another RCA monopoly by innovating an alternative.

One of the basic principles of mass production is interchangeable parts: any screw and any bolt of given dimensions fit one another. That was the goal of RCA's picture tube factory in Lancaster when the engineering staff started manufacturing color CRTs. That is, one could assemble the components of any shadow-mask screen with any picture tube. In 1953, however, Norman Fyler of CBS-Hytron designed a mask that fit flush with the inner faceplate. The phosphors could be deposited directly on the face plate, which simplified assembly significantly. Although this technique required matching and tracking each set of components through production, it became the industry standard.

Throughout the 1950s and early 1960s, before consumers began buying color TVs in significant numbers, inventors offered more radical changes. Most of these designs used differences in voltage, or force, to direct the electron beams on the phosphors. Focus masks used this approach to form electronic lenses within larger shadow-mask holes; Ernest Lawrence's Chromatron used it to control the direction of one electron beam past a mask of vertical wires to strike stripes of red, blue, or green phosphors. Philco's Apple beam-index tube eliminated the mask entirely through complex circuitry and phosphor stripes. RCA's scientists and engineers also tried to eliminate the mask by building an "onion-skin" phosphor, which contained red, green, and blue phosphors illuminated by a single electron beam that changed voltages to penetrate the layers.

None of these overcame RCA's incremental improvements to the shadow mask. In 1964 it began mass-producing rectangular 25-inch color tubes just as Sylvania began using phosphors incorporating rare earths activated by oxide for a brighter red. Martin Royce at RCA's Lancaster, Pennsylvania, factory led the pungent development of rare earths activated by oxysulfide, which by 1968 improved the brightness of red phosphors 45 percent. Seven years later, after the application of a black matrix around the phosphor dots to absorb ambient room light and pigmented phosphors to increase contrast, RCA led other companies in increasing the brightness of their expanding screens fourteen times.

Most television companies continued to draw on RCA science and engineering, but Sony Corporation's Trinitron CRT receivers represented the acme of the Japanese approach to innovation in a mature industry. Under President Masaru Ibuka, Sony had licensed the Chromatron color tube to avoid further RCA licensing fees. The 18,000 receivers made in 1964 with the technology sold poorly in Japan. Three years later, Susumu Yoshida led Senri Miyaoka and Akio Ahkoshi in replacing General Electric Company's single electron gun with three inline cathodes, and the Chromatron's vertical wire grid with a photoetched aperture grille to demonstrate the first Trinitron color picture tube. With ten tubes in hand at the April 1968 press conference, Ibuka promised mass production by October. Sony's harried production engineers and Osaki factory workers fulfilled this "nightmare" by manufacturing 10,000 12-inch receivers by Christmas (*Susumu Yoshida*, 1994, 13).

The Trinitron design simplified the alignment of the electron beams and increased the brightness and resolution. For the next ten years it represented a dramatic improvement over shadow–mask CRTs. While RCA responded with significant improvements, Sony spent up to five times as much on color TV engineering and maintained higher levels of quality control in manufacture. By the mid-1970s, RCA had lost its dominant position in color TV sales to Zenith Corporation. Meanwhile Sony and Matsushita had gained 14 percent of the color TV market from zero in 1968. By 1987, Japanese companies tripled their share and Zenith was the sole American company still making televisions.

FLAT-PANEL DISPLAYS, 1951–1988

The thought of watching television on a flat display runs back to the first conceptions of television in the 1870s. The challenge of making and selling a practical display in any form preoccupied inventors and entrepreneurs for the next eighty years, until David Sarnoff asked RCA's Princeton researchers to "amplify light" as another of his anniversary requests in 1951. The resulting demonstrations disappointed Sarnoff, but the goal remained. So did corporate funding for any member of the staff with an idea on how to make a flat television.

One of them was Jan Rajchman, whose experience in addressing the expanding matrices of ferrite cores for computer memory led him to apply in 1955 for what became U.S. patent number 2,928,894: a color "mural television" using electroluminescent pixels. One challenge with a pixelated screen remained as it had in the 1880s and the 1920s: how do you illuminate each element without affecting adjoining ones, and do so efficiently at broadcast frame rates? Rajchman proposed using "active" matrices to address the individual pixels, as he had for his electromagnetic computer memories, but these required complex and expensive circuitry to store and discharge electrons. Meanwhile others at RCA and the University of Illinois pursued other avenues to a flat video display.

INVENTING LIQUID CRYSTAL DISPLAYS, 1962–1970

Richard Williams, a physical chemist of diverse interests, was one of these people. He not only worked at "a large laboratory for investigating displays" but his work at the David Sarnoff Research Center made him aware of transparent conductive electrodes (Johnstone, 1999, 95). These he joined to liquid crystals, materials that retained a crystalline molecular structure between two melting points. These chemical curiosities had been the subject of episodic research since their discovery in the nineteenth century. Only when George Gray in England published a book on liquid crystals in 1958 did scientists begin examining their properties and potential applications in earnest.

In April 1962 Williams sealed some liquid crystals between two plates of glass partly coated with the electrodes. When he applied 120 degrees centigrade to melt the materials, and a low-voltage current through the electrodes, the liquid crystals under the electrodes turned from clear to opaque, threaded "domains." Turning up the voltage another order of magnitude gave a turbid effect, also scattering light in all directions. His director, Simon Larach, duplicated his work a month later and showed it to senior members of the research center. Williams published two articles on his discovery and filed for a patent that reviewed techniques for matrix addressing of individual elements in a reflective or transmissive video display. His application for a low-power, electro-optic display using liquid crystals issued in 1967 as U.S. patent number 3,322,485.

The chemist had proposed a novel approach to the question of flat displays; the challenge remained in reducing it to practice. However one used liquid crystals for displays, they would have to work at room temperature: Williams's material was liquid only above the boiling point of water. They would also require an integrated matrix of electrodes through which video signals could control the crystals at each pixel; down the DSRC's hallways, Paul Weimer was just beginning to design and fabricate transistors that would eventually address liquid crystal displays (LCDs).

When Williams returned from a research sabbatical in Zürich in 1964, no one had followed up on his proposal. In a company based on electronics, few people had the skills needed to develop electrochemical devices. George Heilmeier was one of them. A talented and ambitious engineer, he had been pursuing other approaches to flat-panel television when he heard about Williams's liquid crystals. Late in 1964, he, technician Louis Zanoni, and chemist Lucian Barton replicated Williams's demonstration and then tried to make a color LCD. Heilmeier added small amounts of a dye to the crystals and Zanoni made up a display the following spring that highlighted the RCA logo in red. This successful demonstration of Heilmeier's guest-host effect suggested to the excited engineer that "wall-sized flat panel color TV was just around the corner—all you had to do was ask us!" (Johnstone, 1999, 97).

In further experiments to find more durable and cooler liquid crystals, Heilmeier rediscovered Williams's turbid electro-optic effect with liquid crystals that the chemists Joel Goldmacher and Joseph Castellano were developing for room-temperature operation. This he called dynamic scattering mode (DSM). Under the higher voltage, the liquid crystals lost their arrangement and deflected light in all directions, giving a flat white appearance. DSM required no polarizing filters that would limit light output to shape the effect; its contrast ratio was 15:1, seven times greater than Williams domains, while it used 10 to 100 times less power.

As a result of the group's work and Heilmeier's advocacy, RCA made LCDs a secret project to begin developing a flat-panel television. More staff joined Heilmeier's group, and Bernard Lechner, who reported to Rajchman in the Displays Laboratory, led another team to figure out how to address each pixel on an LCD television receiver. Over the next three and a half years, while Heilmeier's team discovered two more electro-optic effects, Lechner's team drew on Zanoni's LCD cells to design circuits of Weimer's thin-film transistors (TFTs) to control the reflection or passage of monochrome light at the video broadcast rate of 30 frames per second.

Late in May 1968, an ailing David Sarnoff attended the dress rehearsal for RCA's press conference announcing LCDs. He was no longer well enough to appear in public, but the announcement and demonstrations, held the next day at the Rockefeller Center auditorium gained international attention. Prior to RCA's announcement, liquid crystals served only as crude thermal sensors. Now LCDs appeared to be the solution to flat-panel television.

But how close were they? Over the next three years RCA's researchers spread the word about LCDs and video applications at professional conferences. Television applications awaited far more than the need to improve durability and contrast. It was clear that no one would soon solve the challenge of addressing enough liquid crystal pixels to provide a practical rival display to the 13-inch, monochrome, CRT receiver, which RCA's marketers defined as the minimum for commercialization. Lechner's group remained stuck on its 2×18 LCD that showed off grayscale motion at 30 frames per second; the team never made the 30×40 active-matrix video display it planned.

Inspired by Richard Williams's discovery of a liquid-crystal electro-optical effect in 1962, shown on a microscope slide in the upper right, George Heilmeier led the team that invented and developed the world's first, room-temperature, liquid-crystal displays, demonstrated by RCA in New York in May 1968. David Sarnoff Library

Because of the company's ambivalence over commercializing the technology, RCA's LCD staff dispersed to other projects, start-up companies, or corporations. They joined dozens and then hundreds of chemists, engineers, and physicists in an emerging industry using LCDs for numeric displays, especially in wristwatches, through more versatile materials and a new electro-optic effect. After the press conference, two people began thinking about other ways to modulate light through liquid crystals. Wolfgang Helfrich, who joined Heilmeier's group, and George Fergason at Kent State University in Ohio both examined the helical behavior of nematic liquid crystals. Independently of each other in 1969, both proposed control of light through polarizing filters and the helical twisting of liquid crystals. Application of an electric current straightened the crystal molecules and blocked the passage of light, all for half the power of DSM.

Fergason demonstrated his twisted nematic (TN) approach in December 1969 and published an article with two colleagues a month later in the trade journal *Electro Technology*. Meanwhile, Helfrich could not persuade

Heilmeier to try TN liquid crystals with polarizers in 1969. He joined Hoffman-La Roche in Switzerland in the fall of 1970 and together with Martin Schadt demonstrated a TN LCD that November. The technique was not that impressive—first demonstrations of principle rarely are—but the effect required half the power of DSM for the same 15:1 contrast ratio. With further improvement in the liquid crystals and the circuitry, the TN power consumption dropped to one fifth of DSM with ever-higher contrast ratios. The two filed for a Swiss patent in December, while Fergason deferred his application until the following April.

INVENTING PLASMA DISPLAYS, 1964–1971

While RCA's researchers pursued flat TV displays using liquid crystals, a quite different group invented flat computer displays using plasma, the electronic stimulation of gas into charged atoms, or ions, and electrons. Beginning in 1960, University of Illinois professors Donald Bitzer and H. Gene Slottow had put the ILLIAC computer to work in educational applications. They needed a better interface between people and machine than a teletype printout or punched ticker tape, and one that could retain information on a display. The CRTs traditionally used for computers refreshed the information on-screen from data stored in specialized and expensive vacuum tubes. Bitzer and Slottow conceived an alternative in July 1964 while waiting for their wives to pick them up from work. This was a gas-filled tube similar to neon and fluorescent lighting, only filled with cells of gas to make an electrically addressable plasma. Together with graduate student Robert Willson, they sealed two microscope-slide covers over a cell filled with neon gas. After coating the outside surfaces with vertical and horizontal gold strips of electrodes, they turned them on.

Was one pixel a plasma display? Someone working with binary data, processing ones and zeros, would answer positively. It contained the essential features of plasma displays forty years later. The slide covers served as dielectric, or insulating, layers that limited the alternating current (AC) power to the gas. At the same time their inner surfaces stored some electrons that sustained the neon glow between the pulses of AC. An inadvertent air leak in the epoxy glue sealing the cell added a tiny amount of nitrogen, which turned out to be essential in attaining the desired memory effect. The three men filed for what became U.S. Patent number 3,559,190, issued in 1971 and assigned to the university.

In the meantime, Bitzer and Slottow scaled up to a 4 × 4 matrix-addressed panel that they reported to the American Federation of

Information Processing Societies in 1966, followed by a 16 × 16 display in 1967. That same year Edward Stredde made a three-cell color panel for his master's degree project, using red and green phosphors excited by ultraviolet light from xenon plasma, and the xenon plasma itself to create blue. Even before receiving patents on these technologies, the University of Illinois sold an exclusive license to the glass company Owens-Illinois in 1967. Engineers there took four years to convert the academics' fragile demonstrations into a commercial product, a 12-inch, 512 × 512, monochrome graphics computer display.

The qualities of plasma that loaned themselves to single-color storage in a monitor, however, were then antithetical to the changing brightness and imagery of full-color television. In addition, the further development of CRT monitors, computer memory, and integrated circuitry reduced the original incentive for the technology. While the U.S. Air Force and Navy invested in high-resolution monochrome plasma displays during the 1970s, no American television manufacturer was willing to underwrite the long-term innovations necessary to overcome the technical and price advantages of color CRTs. Solving plasma's problems, which included durability, brightness, and power consumption, fell to the collective resources of Japan's display industry and to American researchers working in entrepreneurial niches.

INVENTING LCD TELEVISION, 1971–1988

Hoffman-La Roche eventually bought Fergason's patent rights for what became the standard approach to LCDs. Meanwhile, Sharp Electronics of Japan bought a license to RCA's LCD patents in 1971 for three million dollars and proceeded to invest $200 million in LCD factories over the next ten years. RCA; AT&T, which investigated LCDs for its videophone in 1969–1970; North American Rockwell; and Hewlett-Packard all abandoned the technology and the industry to the Japanese, whose profits on digital watches and pocket calculators helped offset the costs of further innovation.

By the late 1970s, with the markets for those products and their segmented numeric displays saturated, production engineers tried simple-matrix addressing of TN liquid crystal pixels. In these LCDs, used initially in computers, the columns of conductive electrodes receive an electronic signal of an image simultaneously while the horizontal lines of electrodes receive their information sequentially. As with Nipkow's electromechanical TV systems, however, simple matrices faced physical limits. When the engineers tried to increase the number of lines beyond 128 or the refresh rate beyond thirty-two times a second, the image lost contrast and viewing angle shrank; there was not enough energy or time for the liquid crystals to react. Increasing the voltage, or force, that drove the electronic signal also activated pixels around the intended one.

The solution lay in active-matrix (AM) addressing. In 1968 RCA's Frank Marlowe of Lechner's group proposed using an array of thin-film transistors (TFTs), acting as switches, to control the behavior of each pixel individually. That was easier said than done, and five years passed before T. Peter Brody led a military-funded group at Westinghouse that demonstrated a 6-inch-square, 14,000-pixel, AM-TFT LCD. The Air Force then canceled its $60,000 contract because, as one witness observed, aside from "other technologists, no one else would have been able to see that it had any merit" (Johnstone, 1999, 124). Brody, Fang-Chen Luo, and others persisted against management's reluctance and showed an active-matrix, monochrome, electroluminescent, video display in 1978, but Westinghouse closed the project the following year.

While the frustrated Brody joined Fergason in the American subculture of flat-panel entrepreneurs, Japanese companies pursued commercialization on a scale that no U.S. company would match. Working seven days a week, Morozumi Shinji and Oguchi Koichi at Suwa Seikosha used Heilmeier's guest-host effect to build an AM-TFT wristwatch television in 1982. Instead of displaying an image by reflecting light from the desired liquid-crystal pixels, however, this display transmitted light from a fluorescent light bulb behind the liquid crystals. These acted as pixelated shutters controlled by the video signal. The blue monochrome LCD adorned Roger Moore's wrist in the James Bond film *Octopussy*, and earned further investment. "I spent a lot of money," Morozumi laughed, mentioning a sum of two million dollars. "I'd send the proposal to management, and they'd be like, 'Oh yes, do it,' and they'd sign it easily" (Johnstone, 1999, 126–127).

The first color LCD TV, introduced at the 1983 Society for Information Display meeting in Philadelphia, was a little over two square inches. Thus the viewing area matched that of a 1920s Nipkow-disc display, but in a much smaller container. It showed a 240 × 240-pixel screen using red, green, and blue color filters to control the transmission of light. The "amazed," mostly American audience wanted to know if Seiko's TFT technology was commercially practical (Castellano, 2005, 85). The newly renamed Epson responded by selling its $495, 1-pound, "Epson Elf" pocket TV between 1984 and 1986.

Liquid crystal TVs remained 3-inch novelties bought by the millions, mostly in Japan, until Isamu Washizuka took charge of Sharp Electronics'

Liquid Crystal Division. The material used to make the TFTs had been the limiting factor until Scottish researchers showed that non-crystalline, or amorphous, silicon improved production yields. In 1987 Washizuka directed his engineers to increase the size of Sharp's LCD televisions from three to 14 inches diagonally to achieve a practical size for home video use. Less than two years later, using existing factory equipment, the inventors showed the 4-pound, 1-inch-thick, color LCD TV to Sharp's president, and then the public. Although research groups in Europe, the United States, and Japan all pushed the size of color LCDs using AM TFTs in the mid-1980s, the breakthrough beyond incremental boundaries stimulated Sharp's technical and business staff. "It was a very emotional thing for us, that 14-inch screen," recalled one of the engineers. "And we thought if we can build something as big as that, then maybe the LCD will be able to rival the CRT . . . we were able to dream about future prospects—that was the most important thing" (Johnstone, 1999, 141).

INVENTING A PLATFORM FOR PLASMA TELEVISION, 1972–1987

Other Japanese entrepreneurs and inventors felt the same way about plasma technology. NHK, Japan's broadcast company, started researching high-definition television in the late 1960s and stimulated work in large TV displays. Since the human eye lacked the acuity to discern the differences in resolution on smaller displays beyond a certain distance, viewers would need larger displays, using a technology less bulky than CRTs, to appreciate what NHK intended to offer. As a result, Fujitsu led a troop of Japanese companies in making licensing visits to the University of Illinois laboratory. Plasma also enabled the display of kanji, the Japanese ideograms, stimulating numerous markets wherever alphanumeric displays were used, and therefore a way to underwrite the enormous investment in plasma television.

The first problem to be solved was the issue of gray scale, since plasma pixels were either on or off. Engineers at Hitachi and Mitsubishi resolved this in 1972 by writing and erasing the pixels at higher speeds in a given time span, with incoming video information determining the level of relative brightness in each pixel. Six years later NHK showed a 16-inch color display, provoking the hope that the commercial TV on a wall was not far away.

Isamu Washizuka, head of Sharp Electronics' Liquid Crystal Division, shows off the 14-inch, color, LCD TV that operated at broadcast standard resolution and frame rate, in 1988. David Sarnoff Library, courtesy Sharp Electronics

Nonetheless, well into the 1980s, plasma video displays still paled in comparison with the color picture tube in cost, size, power consumption, and literally in brightness. It remained in 1979 for Fujitsu's staff to maintain the phosphors' luminance by separating them from the cathode with a now-common technique, and Hitachi's engineers to increase the light output by a factor of five in 1984, to 1.5 lumens per watt. Fujitsu and AT&T's Bell Laboratories removed the visible spacing between pixels in the mid-1980s by using three electrodes to address each red, green, or blue subpixel, and separating the electrodes sustaining a pixel and that one providing the video data. Plasma displays also still consumed far more power than CRTs per display element, although Larry Weber and M. B. Wood at the University of Illinois invented an electronic circuit that reduced the power needed to sustain the plasma cells by over 100 watts. That breakthrough, combined with the American military supplier Photonics' demonstration of a 59-inch, 2,048 × 2,048, high-definition monochrome display in 1987, demonstrated the commercial prospects for plasma television as governments around the world began approving standards for high-definition TV.

LOOKING AHEAD

In the early 1960s, David Sarnoff anticipated a global information culture by the end of the century where people would exchange "thoughts, documents, and data through desk instruments and a color-TV screen on the wall." Alternately, when "equipped with miniature TV transmitter-receivers [they] will communicate with one another" (Sarnoff, 1968, 163, 287). Technological developments described in this chapter begin to make the culmination of that vision clear. But Sarnoff never imagined its fruition would not involve RCA, whose board of directors agreed to its purchase by General Electric Company (GE) in December 1985. Two years later GE's chief executive, Jack Welch, traded the Consumer Electronics Division to the French company Thomson for its medical imaging unit and cash, and donated the David Sarnoff Research Center to SRI International, a nonprofit research organization, in return for a substantial tax deduction. The DSRC and Thomson/RCA continued their technical traditions in contributing to the final frontier of television as a discrete technological system with unique components. But the death of television took place at the hands and minds of people, companies, industries, and governments around the world, and it is to that demise we turn in the final chapter.

6

The Digital Generation
and the End of Television

◆

LOOKING BACK AT 2006

By the end of 2006, the technology of television had changed beyond recognition—if not overnight, then over the previous five years. One Thursday night in December, the Princeton/Central Jersey joint chapter of the Association for Computing Machinery and the IEEE Computer Society hosted an evening devoted to digital high-definition television (HDTV). Dr. Michael Isnardi of Sarnoff Corporation plugged a small cabled device into his laptop computer on the company's auditorium stage. The other end connected to a cable that ran up to an antenna on the roof that received high-definition television signals broadcast by New Jersey Network and stations in Philadelphia. Isnardi could have watched HDTV on his laptop, but instead he shared the programs by sending the signal through another cable to a liquid crystal display (LCD) projector that shone the bright, sharp imagery on a 10-foot screen.

Elsewhere, around the world via cellphones and digital cameras; the Internet and local WiFi wireless nodes; and fiber-optic telephone and cable networks, corporations and individuals operated their own television systems and created their own, often transient, networks, content, and audiences. People spent more time online than in front of a TV and the younger they were, the more that was the case. They installed webcams on their Web sites, edited digital video files on their computers, and uploaded and downloaded homemade and corporate videos to and from YouTube, Jumpcut, and countless other sites and blogs. Some 150,000 people watched an appearance by Jon Stewart live on the CNN talk show *Crossfire*, and ten million more watched the recording on Web sites like Bittorrent, iFilm, and Google Video.

Commercial broadcast television remained an imposing cultural and economic force. Yet its executives recognized that their declining audiences and stagnating net profits of $20 billion in advertising revenue arose from more than traditional rivals for eyeballs: cable and satellite TV services and VHS cassettes or the digital videodiscs (DVDs) introduced in 1997. The new competition arose out of the digital domain of personal computers, microprocessors, and electronic games, connected to the decentralized network of the World Wide Web that Tim Berners Lee introduced to the Internet in 1993.

To survive, much less expand, the traditional networks and video delivery services began diversifying into the exploding panoply of online and wireless media choices and program formats. Rather than insist that the audiences justifying ad rates come to the networks and their broadcast channels for programs, the networks began selling digital files for downloading onto a variety of media. Now younger audiences could watch and increasingly interact with shows and sports events at their convenience on their laptops, Macintosh Ipods, cellphones, or other digital, portable, display devices. ABC put the first broadcast network program, *World News Now*, on the Internet, in 1994. In December 2005 NBC Universal began selling downloads of its popular show, *The Office*, on Macintosh's iTunes Web site and announced it would offer ten mini-episodes produced exclusively for its Web site in the summer of 2007.

Just like the NBC managers producing television for the regular schedule that David Sarnoff announced in 1939, the production staffs in the new century struggled with tiny budgets. Their problem was the same chicken and egg of sponsors and viewers: why underwrite a show that few people are watching? Now, however, digitization meant there would always be programmers providing content to skip—or select. Nearly 10 percent of households already used digital video recorders like TiVo to watch programs on their schedule—and to skip advertisements in the programs. Other Americans watched older episodes of popular shows on DVDs, which contributed to a growing sense that viewers should not have to endure the unsolicited ads that paid for programs and earned broadcasters their 40 percent profit margins.

The fact that this all occurred at the turn of the millennium was coincidental, but the beginning of the twenty-first century marked the deathwatch for television as two generations of people had experienced it. What do we mean by *death*? In this case, the concept of a population all passively watching the same professionally produced program simultaneously—much less on a common type of display—at a government-approved broadcast standard, was fast becoming a phenomenon of the past, resurrected for sports championships and national disasters. For children growing up in a digital and internet-worked culture, television no longer existed as a separate medium for the distribution of information. Instead they, web entrepreneurs, phone companies, Hollywood, publishers, the computer industry, and everyone else blended video content with audio, text, and data in a seamless, four-dimensional web. Anyone could create, edit, transmit, receive, or interact with anyone else's material, at a variety of resolutions and bandwidths, at their convenience. This chapter reviews how people changed TV from an analog to a digital standard in the years before 2000, and simultaneously expanded the versatility of displays, and by doing so ended television's life as a discrete technological system.

WHY CHANGE TELEVISION?

When does a nation change its broadcast standards? Since television signals pass through the atmosphere above a country in the electromagnetic spectrum, the national government asserts the right and responsibility of deciding how that spectrum is allocated. In the United States, the government through its Federal Communications Commission typically leaves it to the broadcast industry or the military to propose uses of the spectrum and standards for that use. A new use implies that a current use will be discarded or reallocated on another bandwidth. It also raises the tension between what the broadcast industry and the government believe is the fairest use for the people. The former, after all, regard them as consumers paying for products; the latter view them as public citizens whose trust resides in the FCC's judgment.

To change a standard also implies a change in system, from pickup to transmission format to signal propagation to receiver. Those technologies involve not only the government but industries full of people with a vested interest in maintaining the status quo. Without an innovative monopoly like RCA under David Sarnoff earlier in the twentieth century, the United States had little incentive to change its broadcast television system in the late 1980s.

Yet innovators around the world had improved significantly the technologies that comprised TV. Inventors and entrepreneurs in the semiconductor electronics industry had been improving the reliability, speed, size, and cost of integrated circuits (ICs) since their invention in 1958, and the cost of microprocessors—computers on chips of silicon—since 1971. For television engineers, ICs increased the ability to process and control video signals, among other applications. The improvements discussed in the previous chapter for cameras, recording, and displays continued, expanding the resolution capabilities of image capture and display.

Technical professionals in video processing as well as equipment manufacture agreed on the advantages of raising the definition of television imagery. To move from 525 lines to about 1,000 lines per frame would approximate the two-million-pixel resolution of 35mm motion-picture film and improve the picture. Inventors also needed to widen the video frame to match the dimensions of modern motion pictures. The original aspect ratio for electronic television displays of 4:3, or 1.33:1, suited both the Hollywood film standard of the 1930s and the manufacture of early cathode-ray tubes (CRTs). The closer the proportions of the display frame to a circle, the less stress there would be on the glass envelope containing the tube's vacuum. Since the early 1950s, however, when Hollywood studios responded to the popularity of television by offering wide-screen formats, most movies had appeared in a ratio of 1.85:1. Broadcasters then showed only the center of the movie frames, or "panned and scanned" the film, focusing on what appeared to be important in the imagery.

The question was, should the move to high definition be analog or digital? The French and the English had initiated work on analog HDTV in the 1940s. In response to a government commission recommendation three years before, EMI demonstrated 1,001-line monochrome video in 1948. The transmission used 25 MHz of bandwidth; the French implemented their 14-MHz, 819-line, monochrome standard that year and retained it until 1986. With no or few commercial TV stations and channels, the Europeans and the Japanese enjoyed more flexibility in use of bandwidth. In any case the high resolution was lost on the small screens available to most viewers, as they watched TV from a distance that sapped program content of any immediacy. A true HDTV system would require a larger display, along with stereo sound, in which to immerse and engage home viewers.

THE ORIGINS OF HIGH-DEFINITION TELEVISION, 1969–1986

Applying digital techniques to entertain and inform the public was impractical in the 1960s. Thus Japan's national broadcast network, NHK, began research in 1969 to match the resolution of film in television through analog techniques. Its engineers became increasingly aware of the flaws in the

American NTSC system as they worked with higher quality cameras and monitors. To watch television at a distance where the resolution appeared sharp removed the viewer physically and emotionally from the screen. The goal of a new, Japanese-based system should be to "appeal to a higher level of psychological sensation and emotion by transmitting highly intellectual information with detailed characters and graphics" (Brinkley, 1997, 13).

By 1972 NHK's engineers under Masao Sugimoto had performed numerous studies of the physics of visual perception to discover what would be required to simulate a sense of reality on a large-screen television. They settled on transmitting 1,125 lines, sixty times a second, with a 16:9 aspect ratio. This quantity of information required over 20 MHz of bandwidth, over three times that used by existing channels, but NHK anticipated using a single satellite channel transmitted to home dish antennas using gigahertz (GHz) frequencies.

That March the Japanese government also asked the Commité Consultatif Internationale de la Radio (CCIR), the international standards organization, to begin considering HDTV standards. The main charge to Study Group 12 was to define analog and digital parameters for an HDTV standard. NHK's system provided a template for a quality such that someone viewing the imagery at three times the height of the display would perceive the same resolution, color, motion, depth of field, sounds, and immersion in the scene as in real life. It also promised Japanese companies control of the new HDTV industries based on the head start given by their relationships with NHK.

In 1978 NHK began demonstrating its system, called MUSE after the form of compression it used, to fit the 20 MHz of information into the 8 MHz of bandwidth available through its satellite receiver and transmitter. When NHK offered MUSE in 1986 to the European nations for possible adoption, with equipment patented by Sony and other Japanese companies, they balked. As with the earlier debates over an international color TV standard, their governments and communications industries believed it economically vital to retain the ability and incentive to innovate high technologies domestically. Their response between 1986 and 1990 was another analog high-definition system, MAC, based on a billion-dollar investment by 30 laboratories and companies.

ANALOG HDTV IN THE UNITED STATES, 1986–1990

Where was the United States? While Japan and Europe moved ahead with high-definition standards under government subsidies and support, its government distanced itself from the disintegrating consumer electronics industry. Republican free-market administrations left it to the relevant industries and their Wall Street investors to decide what to offer consumers as new formats for information and entertainment. In an increasingly deregulated economy, markets shaped by tax policies and the uncoordinated decisions of consumers, investors, executives, and inventors would determine the evolution of consumer electronics, and which companies would thrive or disappear. Alerted by Hollywood to the value of programs and movies exported overseas on a common HDTV standard, the State Department signed off on NHK's proposal. Japan already dominated American consumer electronics market, and an American standard would not bring back domestic factories and employment. Thus, when the department heard in 1985 that RCA had begun mulling its own HDTV system, a bureaucrat warned its president, Robert Frederick, that a lack of cooperation on a global, NHK-based, standard would have "significant adverse consequences to U.S. trade and information interests" (Brinkley, 1997, 56).

At the industrial level, the National Association of Broadcasters (NAB) only began lobbying the Federal Communications Commission (FCC) in 1986 for a high-definition standard in order to protect unused ultrahigh frequency (UHF) channels from reassignment to mobile radio applications. Executives at the major TV networks wanted a new American system of broadcasting only if it remained compatible with current technologies. In other words, new cameras, transmitters, displays, and related equipment had to work with current standards and whatever new one the FCC approved. This would save them and their affiliates money during the transition and made the prospect of a truly high-definition system unlikely.

To support this approach, the networks and leading American broadcast equipment manufacturers funded a Center for Advanced Television Studies (CATS) at the Massachusetts Institute of Technology (MIT) in 1983. They hired William Schreiber, the respected academic authority on broadcast engineering, to head the center as an honest broker. He had no industrial biases and therefore he also had little influence. As time passed Schreiber became less enthusiastic with his charge: what was the point of training the best American students to develop only a better system?

In 1986 Schreiber moved CATS to Professor Nicholas Negroponte's Media Lab. Negroponte based the Lab's value on its ability to research technologies that could become commercially available in five to ten years, and one of those was digital television, not an analog, high-definition system. Another academic outsider crossing industrial and technological boundaries, Negroponte was deeply versed in digital interactivity and the networks available to personal computers. "HDTV is a lot of hot air," he asserted regularly in the print media. The FCC should only approve a digital system if it enabled "the complete integration of computers into television sets" (Brinkley, 1997, 94).

Meanwhile RCA's David Sarnoff Research Center, which General Electric was turning into an independent company, had proposed an enhanced, widescreen television system called Advanced Compatible Television (ACTV). Similar in technique to MUSE and MAC, it permitted consumers with new 16:9 displays to view a full picture while those with sets with the older 4:3 ratio could see the center of the image. The NAB embraced ACTV because the DSRC staff claimed it would cost only a few hundred thousand dollars to upgrade equipment per station, compared to the estimated $10 million that incompatible HDTV would cost. The DSRC also promised to increase its bandwidth demands to a second channel with a next-generation version, to take place in an unspecified and therefore inexpensive future. The Dutch company Philips and Thomson of France, which had obtained RCA's Consumer Electronics Division, allied with the DSRC because it aligned well with European interests. They could also sell TV receivers with the 16:9 format now, and another, higher-definition format in the future.

Despite ACTV's advantages, after the DSRC demonstrated it on the fiftieth anniversary of David Sarnoff's introduction of RCA's TV system in 1939, it had become irrelevant by 1990. More members of Congress had seen the NHK demonstrations, and the FCC had gone through two more chairmen. The latest, Alfred Sikes, supported by Richard Wiley of the FCC's Advisory Committee, had approved an international competition for the best HDTV system in 1988–1989. Sikes was unhappy with the indifferent improvements of the seven finalists. After all, whoever won would receive royalties on the circuits designed for their HDTV system from anyone who made broadcast and display equipment in the United States. To earn that right, the winner should offer true high-definition. Wiley had visited Negroponte's Media Lab and become persuaded of the advantages of a digital format. In March 1990, then, Sikes issued a new requirement doubling the current television resolution, in the standard broadcast channel of 6 MHz. In addition, the system had to simulcast HDTV programming on a separate channel in traditional NTSC format. That would enable a gradual transition to HDTV while the public bought HDTV receivers.

Sikes and Wiley's challenge virtually demanded a digital system, where the signal could be compressed to give the necessary resolution within the same size channel approved for electronic monochrome television in 1941. That, not the camera or the display, was the primary challenge in making the system work. A standard frame of the 35mm film used to produce 85 percent of prime-time programming provided the equivalent of two million pixels of color information; television required 30 frames a second. That amounted to the transmission of 60 megapixels a second as 60 megabits, in a 6-MHz channel. Contestants would have to remove 90 percent of the signal and restore it at the other end, in real time. No wonder that CBS's vice president for technology, Joe Flaherty, predicted that "We'll have digital television the same day we have an antigravity machine" (Brinkley, 1997, 94).

THE ORIGINS OF DIGITAL TELEVISION, 1956–1988

A digital format enables precise control of the processing of electronic signals as they pass through the electronic and thermal noise of a television system. The stimulation of electronic activity in any device generates energy via the movement of electrons. Some of the electrons move in ways not desired by the designers, and they and the heat given off by all motion interfere with what people want electronic equipment to do. Electromagnetic fields generated by other electronic equipment, or solar and other electromagnetic phenomena in the broadcast spectrum, also affect a wireless signal.

These noises distort the shape of the continuous analog wave of electronic broadcasts. Many engineers make a living understanding, compensating for, or overcoming analog noise in electronic circuits. If instead they could quantify, or convert, electronic representations of the real world into encoded streams of bits, the ones and zeros of binary counting, then they could write programs to maintain the digital pattern cleanly throughout the noisy conversion and transmission process between capture and display.

The first digital television system originated between 1956 and 1961 with an ex-RCA engineer who responded to a series of military and National Security Agency contracts for encrypted video communications. Richard Webb converted the analog brightness data from a monochrome video signal into a binary bitstream encrypted by the military's new digital computers. Drawing on a modulation technique Frank deJager of Philips used for audio recordings, Webb and his company's engineers in Colorado used it to encode the relative, rather than absolute, brightness of each pixel, saving significant bandwidth and generating a single channel of data for transmission. They implemented this approach for President Dwight Eisenhower, who reportedly said that "if [Secretary of State] Allen Dulles ever calls me to push the big red button on Russia, I want to see the expression on his face" (Webb, 2005, 144). Two years later, the White House had the world's first digital TV station, which exchanged square-frame, 405-line, video images with receiver systems at the Central Intelligence Agency and Camp David via 6-MHz microwave antennas. The White House system remained in place until 1979, an example of what one could accomplish with ingenuity, research, and a huge budget.

By the late 1970s the price of digital electronic circuits and computers had dropped and the sampling rates had improved so that broadcast engineers began using them to generate acceptable approximations of analog measurements. Digital techniques found their way into Bosch Fernseh's film-to-video converters, or telecines, in 1979; RCA's microprocessor-controlled TK-47 studio camera; and the British Independent Broadcast Authority's demonstration with Bosch Fernseh in 1978 of a digital video-tape recorder (VTR). Ampex in California and Sony followed the next year with the first prototype VTRs, and Sony began selling its D-1 digital videotape and recorders in 1983. Entrepreneurs, scientists, and engineers in broadcasting who tracked developments in the semiconductor industry and the conformity of transistor density on ICs to Moore's Law could anticipate nothing but more powerful digital tools long into the future.

CONVERTING TO DIGITAL HDTV, 1988–1990

Flaherty was one of the biggest proponents of HDTV and remained open to innovations, but his comment on its impossibility reflected the culture of most managers steeped in a mature and profitable system. They were aware of the improvements in electronic processing power and digital developments around the periphery of the system, but they were not innovators or entrepreneurs. The radical changes took place among a younger generation of engineers at small or struggling companies willing to take a risk to survive or working in one of the industries spawned from broadcast television, cable, or satellite.

Among those engineers was the staff at Zenith Electronic Corporation on the plains outside Chicago. The company was losing millions of dollars annually even as it moved TV receiver production out of the United States to Mexico. One of its smaller businesses was in the cable converter boxes that turned the video signals from a cable TV provider into the frequency for channel 3 or 4 on a home receiver. Rich Citta led the team that digitized part of the television signal, which eased encryption for the cable provider; shrank and clarified the signal; and increased the number of channels the provider could deliver on the same cable lines. While recovering from a broken ankle in 1988, Citta realized that Zenith could enter the HDTV competition with a 787-line television image in one partly digital channel and simulcast the NTSC version on another.

Zenith's chief executive, Jerry Pearlman, and Citta's boss Wayne Luplow gave his concept full support. The company had lost $52 million in the previous year and HDTV was its last hope for survival. Ironically, the positive publicity on Zenith's decision led the U.S. government to prevent it from selling its unprofitable consumer electronic business to foreign bidders. Instead Pearlman had to sell its successful personal computer division to offset the losses and underwrite more research and development on Citta's HDTV system.

In San Diego, Jerry Heller of the VideoCipher division of M/A-Com began evaluating HDTV from the perspective of its effect on the satellite television industry. Because of the size of the MUSE channel, it would render obsolete the millions of satellite dishes in yards across the country. The division already suffered from the determined owners of dishes who fought the cable industry's efforts to make them pay for reception of satellite transmissions of cable channels (see Chapter 5). Heller, fellow MIT alumnus Lee Paik, and other engineers had first scrambled the satellite signal digitally and then built the VideoCipher II set-top box so that satellite-dish owners could pay to descramble the signals. By 1987 private inventors had developed and had manufactured an IC to evade VideoCipher's encoding. Even if VideoCipher blocked the hackers, HDTV would drive them out of business. When Paik reviewed the HDTV proposals and concluded that all of them would require bigger dishes, Heller pushed him to imagine a system that would not.

The obvious solution for a company working outside the broadcast industry's traditions and experienced in digital modems and signal scramblers was a digital system. Paik fit all that binary information in a limited bandwidth by compressing it. Each frame of video contains pixels whose color and brightness information remains static over successive frames, and the repetition can be discarded in transmission. Other pixels' information moves in a particular direction. With a digital storage buffer, algorithms for repeating the behavior of bits mathematically, and exploitation of the limits of human visual perception, Paik and another MIT graduate, Ed Krauss, began months of writing computer code in the basement lab to eliminate data and finesse selected frames containing scene changes with lower resolution because there was no place to put the bits. Using sample film clips from the films *Top Gun* and *Indiana Jones and the Temple of Doom*, they fit a digital HDTV signal into the bandwidth of a satellite channel by early 1989.

What was the market for such a breakthrough? The division, now owned by General Instruments (GI), had begun under the assumption that the Japanese MUSE system and HD television receivers would soon be available. That was no longer the case, and Heller told Paik to compress the signal again twofold to fit it in a cable or broadcast system. This took him and Krauss well into 1990. In the process they fit four digitized NTSC channels into a standard cable channel. This was an application with immediate commercial appeal.

GI took their digital cable system to a trade show where they attracted the attention of the DSRC and Flaherty. The visits by the east coast executives to San Diego echoed trips to Philo Farnsworth's San Francisco lab by their ancestral peers some sixty years before. While the ex-RCA representatives failed to ally with Heller and GI, Flaherty gained their trust and in turn urged them to join the HDTV race by June 1, 1990. Heller and Paik flew to Washington, DC, to meet Wiley and Sikes by that hot, humid Friday, deliver their check for $175,000, and explain their Digicipher approach. The FCC chairman was overjoyed: "That's just what I said would happen!" (Brinkley, 1997, 141).

COMPETING FOR AN HDTV STANDARD, 1990–1992

GI's system was rudimentary, virtually vaporware untested for broadcasting. But as the only fully digital system it changed the dynamics of the competition. It crushed the Japanese who had invested half a billion dollars developing the analog MUSE system and its hardware. For Glenn Reitmeier, a rising star at the DSRC who had pursued digital video technologies since 1977 under Curt Carlson, the news established him as Sarnoff's HDTV champion now that ACTV was obsolete. Zenith had to accept that its hybrid system was also defunct. At MIT, where Professor Jae Lim was trying to raise funds to test Schreiber's concepts for a hybrid HD system, the news that his MIT classmate and friend Paik never revealed directly to him was stunning.

Although the conditions of the race had changed, the finish line where system testing would take place had not: eight weeks for each entry from mid-1991 to early 1992. The Test Center in Alexandria, Virginia, awaited with dozens of equipment racks for checking resistance to all manners of electrical interference; computer and cable setups for co-channel interference at every potential station assignment in the United States; and a 65-inch Hitachi HD projection receiver on which a group of "golden eyes"—expert viewers—would watch how each system responded to a special HD tape of moving objects, contrast changes, and random motion.

As a result Zenith returned to a prior relationship with AT&T Laboratories in Summit, New Jersey, where the latter's huge technical staff had a deserved reputation but no experience in broadcast engineering. General Instruments, needing help with digital transmission, allied with MIT, where Lim needed money and offered seven top graduate students in exchange. The DSRC continued its relationship with Philips Laboratories in Briarcliff Manor, New York. In the months leading up to each group's test, engineers spent 10, 12, 16, and more hours a day, seven days a week, for months on end, in windowless rooms and cubicles, designing, assembling, testing, and debugging the hardware and software that had to operate and interact in a real-world environment. In the monochrome standards evaluation in 1940–1941, RCA's engineers had been testing and refining their system for ten years. In the color standards contest in 1949–1950, CBS's staff showed off a known technology, but digital HDTV engineers faced a challenge similar to that of RCA's electronic color team: showing on short notice a system that could do what had never been done before. The difference with digital was that if the signal failed to come through, there was no signal at all, leaving engineers to guess where the problem lay.

Most remarkably, GI's Woo Paik spent the first ten months of 1991 building the transmission system. GI followed on by assembling two racks of equipment with three engineers and astonishing the manager of the center, Peter Fannon, who had already endured tests of ACTV and NHK's Narrow MUSE system. "Hot damn! Now we're getting somewhere. *This* is HDTV!" (Brinkley, 1997, 177). GI's corporate representative, Bob Rast, exploited the success by arranging GI's system broadcast to the U.S. Capitol from WETA in Virginia that had the desired effect on invited press and FCC commissioners. "I love it," said Dick Wiley.

The tests proceeded with strict rules governing changes to the systems installed at the center. Each team stretched interpretations, particularly Zenith-AT&T and DSRC-Philips. Far more than GI, they started from scratch and simply lacked the time to create working systems. Zenith and AT&T's team barely trusted one another because of the different contexts from which each worked: a dying consumer electronics company and the greatest research lab in the world.

The Sarnoff and Philips groups cooperated more but failed to get a signal for months. Reitmeier and Philips's Carlo Basile used the new international standard for video compression, MPEG-2, and packetized the signal with layers of additional information. Coming from a background in computer science, Reitmeier understood the potential for making the digital bitstream resemble data sent over computer networks. There, each packet of data referred to its destination and contents, which could include any sort of digital data, not just video. The complexity of ambition suffered under the pressure of the deadline, which the DSRC managed to extend two weeks. Their team arrived in Alexandria with fourteen racks of equipment and once the system started working, it showed imagery as good as GI's.

At the end of the tests, all of the systems showed problems that the contestants blamed on "implementation" at the test center rather than errors in design. This meant that whatever the grades of the golden eyes, the engineers could quickly correct the problems. All proceeded to stage follow-on demonstrations, further promoting the imminent approach of an HDTV standard, and all the new equipment and receivers that its commercial diffusion required.

As dramatic as the images were, however, neither the broadcast industry nor the public showed much interest in high-definition. For network and station owners, HDTV required millions of dollars in equipment and new techniques with no corresponding increase in advertising revenue. In fact they could anticipate losing even more money broadcasting HDTV to tiny audiences of "early adopters," those wealthy and enthusiastic enough to buy the first, $10,000, HDTVs. Surveys showed that for most viewers, high-definition television was no more important than high-fidelity audio had been to radio listeners in earlier generations.

In addition, during 1992 Negroponte had joined forces with Peter Liebhold of Apple Computer to lobby the FCC to redefine the future standard as "open architecture." Unlimited by patent licensing restrictions, anyone would be able to improve on and add options in software programs as well as equipment. High-definition was irrelevant; now that television was digital, it could join in the increasingly interactive, networked world of digital information on home, industrial, and government computers. While agreeable to an academic committed to the free exchange of information and a computer industry that had invested nothing in the digital broadcast standard, this approach gained no support from the consumer electronics industry.

COOPERATING FOR DIGITAL TV STANDARDS, 1993–1996

When Wiley, who now chaired the FCC, announced early in 1993 that all of the systems were flawed enough to mandate expensive retesting, he sparked a revolt and a suggestion by Donald Rumsfeld, the new head of General Instruments. Why not combine the best aspects of each contestant's entry?

The technical leaders of the HDTV Grand Alliance pose in 1995 before their encoding and processing equipment in the David Sarnoff Research Center's Field Lab in Princeton, New Jersey. Standing left to right: Wayne Luplow, Zenith Electronics; Glenn Reitmeier, David Sarnoff Research Center; Robert Rast, General Instruments; Terry Smith, David Sarnoff Research Center. Seated left to right: Ralph Cerbone, AT&T; Jae Lim, MIT; and Aldo Cugnini, Philips. David Sarnoff Library, courtesy Glenn Reitmeier

Rumsfeld's compromise, or the "Grand Alliance" as Wiley termed it, took another seven months to sort out profit sharing, technology assignments, and technology choices.

One choice involved the use of interlaced versus progressive scanning. The former was another legacy of the 1930s, when TV transmission systems lacked the capacity to transmit 30 full frames of video within the six megacycle or smaller channels. Used around the world in broadcasting, interlace techniques had been ignored by computer-monitor manufacturers who had no bandwidth issues. They used progressive scanning, which eliminated the artifacts generated by aligning two alternating lines that scanned motion at different moments in time. Showing interlaced video on computer monitors generated additional flaws. For the sake of a convergent future using

720 lines, Jae Lim of MIT held out for the abandonment of interlace, and was supported by AT&T and computer industry lobbyists. The broadcast traditionalists were equally adamant about retaining the higher scan-rate option of 1,080 lines. Wiley arranged for agreement on an eventual migration to a progressive 1,080-line scan.

Six years after the National Association of Broadcasters started the political process to promote an HDTV standard, President Bill Clinton and Vice President Al Gore moved into the White House in 1993. Gore had put considerable effort in the U.S. Senate into legislation sponsoring the development and expansion of the Internet beyond its military-academic confines, and now he began to tout the digital "information superhighway" that all Americans would use. Neither he nor new FCC chairman Reed Hundt showed great interest in the HDTV standard. When Wiley led a demonstration for Hundt, the chairman was unimpressed, given the broadcast industry's inability to show what else could be done with "the telecomputer side of it. . . . It's not enough to say that this is just a pretty picture" (Brinkley, 1997, 302).

Since the broadcasters agreed with Hundt in order to abandon HDTV for digital broadcast data services, it fell to the Grand Alliance to defend it. Over the course of 1994, the Clinton administration's interest in auctioning unused electromagnetic spectrum and in low-definition, data-based, applications of digital TV gave its members tools to pressure the NAB to show more interest in HDTV. By early 1995, the FCC had raised nearly eight billion dollars through spectrum auctions, leading many in Congress to ask why broadcasters were entitled to two free channels in a digital transition without bidding for them.

At the NAB convention in Las Vegas that spring, industry leaders reversed their earlier rejection of HDTV, having calculated that the necessary equipment was cheaper than bidding on spectrum. In addition, the debut of the Thomson-DSRC-Hughes Aircraft DirecTV digital satellite service that year was becoming the fastest-selling consumer electronic system in history, despite the $1,000 cost of the 18-inch dish antennas. Maybe consumers would pay for the $2,000 HDTV receivers that Zenith promised for 1996. And, at the convention, the most attractive and impressive demonstration of digital television was not the Alliance's examples of interactive advertising or email platforms. Beyond the fact that Tim Berners-Lee had invented the World Wide Web system for the Internet only the year before and few understood its applications, the sheer physicality of the HD monitor's images of a basketball game and dramatic landscapes stunned passersby: "Boy, that was good. I've never seen anything like that" (Brinkley, 1997, 344).

When the FCC met under Hundt to consider the question of the digital television standard in 1995, the broadcasters lined up with promises to transmit HDTV. The NAB pressured congressional representatives successfully, once again using the threat to local free broadcasting to justify retention of a second TV channel during the transition to a new standard. In November Wiley convened the FCC's advisory committee on HDTV and gained unanimous approval of the Grand Alliance's array of interlace and progressive scanning standards at multiple line densities.

After eight years, or twice as long as the FCC spent to establish the color TV standard between 1949 and 1953, and $500 million invested in the FCC's initiative, however, Hundt questioned the need to change to a digital standard. Hundt's fellow commissioners did not share his concerns about giving away bandwidth instead of auctioning it. While he waffled, in June 1996 WRAL in Raleigh, North Carolina, became the first commercial station to broadcast digital high-definition content, although no one could watch because no one in the area had a digital decoder for the signal. NBC affiliate WRC followed with the first live HD broadcast that its engineers admired on the one high-definition monitor in the DC, area.

Finally, on Christmas Eve, 1996, Hundt permitted the commission to vote on a standard. The FCC approved the Wiley committee's eighteen digital formats, including standard, enhanced, and high-definition options with progressive and interlaced scanning, with no guidance on how to apply them. These included the gold standard of 1,080 lines progressively scanned sixty times per second. On December 31, 2006, the nation's television stations would give up their analog channels and broadcast only digital signals, if 85 percent of the nation's households contained television receivers compatible with digital transmission.

Why were there so many standards? The commissioners based their decision partly on indecision by the parties involved, and partly by the assumption that improvements in microprocessors would ease conversions between broadcast, PC, and motion picture scanning and frame rates. They also assumed that market forces—that is, entrepreneurs and inventors seeing an opportunity in a market defined by government edict—would generate the necessary technologies at desirable prices by the time of transition from analog to all-digital TV broadcasting. The following year the commission allotted one digital channel for each TV station and ordered stations in the ten largest markets affiliated with the ABC, CBS, Fox, and NBC networks to start transmitting DTV by May 1, 1999. The affiliates in the next twenty largest markets were to begin broadcasting digitally six months later.

DTV AFTERMATH, 1999–2006

The expectation of a ten-year transition from analog to digital broadcasting was optimistic in a number of ways beyond those of faith in markets. The FCC and industries involved assumed that an average of ten percent of TV owners bought new receivers annually. They also accepted uncritically the various logarithmic curves that industry analyst forecast for the rapid adoption of digital TVs as the price of receivers dropped within that span.

Several factors overrode these assumptions. One was the traditional chicken-and-egg problem of stimulating an expanding number of users and suppliers. Why buy HDTVs when there is nothing to watch? Why pay for HDTV broadcasts when no one can watch them? The 1,600 broadcast stations reluctantly adhered to the timetable for buying and learning to use the digital cameras, computers, transmitters, and other equipment in the major television markets. As with the innovations of electronic monochrome and color television, however, station owners could not charge sponsors for eyeballs that could not see the advertising that underwrote programming.

The fact that some 85 percent of households subscribed and received their television programming via cable or satellite, rather than for free over the airwaves, slowed home adoption enormously. Those industries delayed the expense of integration with digital or high-definition formats until after the turn of the millennium, in part to compete with the enormous installation of fiber-optic networks during the speculative Internet-investment bubble of 1997–2000. In 2002, only one percent of American households could receive DTV. By 2006, about 60 percent of cable subscribers could receive HD channels, and DirecTV had launched the first two satellites for high-definition coverage. These transmitted on the spacious K_a band of electromagnetic spectrum, which covered over three-fifths of American households.

Anecdotal reports indicated, however, that broadcast HDTV endured a variety of problems in transmission and display. The FCC assigned the new channels in the UHF section of the broadcast spectrum. This meant that the television station transmitters needed much more power to radiate as far as the VHF signals in use since 1941. It also meant the signals were much more directional. Household therefore required larger, usually outdoor, antennas, just as all television receivers needed antennas in the years after World War II, and tuning to within a degree or two of the signal's direction. That HD channel then needed conversion into an uncompressed signal that could be played back in a format suitable for an analog, digital, or HD-compatible television, or on a computer.

How many Americans would return to their rooftops to install antennas as their grandparents had for early electronic TV? How many had bought and installed all the components to display the signal correctly? Surveys indicated that consumer interest in watching HDTV on their new large-screen, HD-ready receivers had declined 25 to 50 percent. Instead people bought HD displays to watch higher-definition DVDs and increasing amounts of HD cable programming.

The profusion of digital broadcast standards, delivery systems, recording media and displays within the television industry had the effect of increasing choice beyond rational bounds. As choices manifested themselves in proliferating components, cables, connectors, and programming requirements, consumers concluded that the benefits of educating themselves did not justify the costs in time and physical complexity. They chose essentially not to choose.

Thus, less than two in five people understood the distinction between the term "digital" television standards and the subset of HD formats within it, a consequence of broadcasters finessing the two terms. Seventeen percent of viewers thought they were watching HDTV when in fact they were not. The slipshod sales practices of many retailers contributed to confusion over the three levels of FCC-approved digital TV, as did the readiness of any particular television set to receive and display digital cable, much less broadcast HD, without a converter box. When consumers bought their new receivers they rarely understood the adjustments necessary between the multitude of standards and video sources. Finally, tens of millions of Americans, distanced from the media publicity focusing on affluent, early-adopting, consumers, used the same, durable, analog TV receivers far longer than the decade assumed for them.

Thus, ten years after the approval of the HDTV standard, the only people enjoying free, local, HDTV broadcasting were those affluent enough to hire someone to install and tune the necessary antenna system, or those determined enough to do it themselves. They also had to find out which TV stations in a given area broadcast programs in HD, which required visiting a Web site to identify the programming. Descendants of Nipkow-disc or early electronic television hobbyists, these HD pioneers comprised three percent of the broadcast audience. Where the federal government declined to approve a television standard for the enthusiasts in the late 1920s, however, in the 21st century it avoided a debate on why the FCC continued to reserve so much broadcast spectrum for an insignificant percentage of the public. Instead, in February 2006 Congress passed another law deferring the digital transition date to February 2009.

THE GROWTH OF DISPLAYS, 1988–2006

The transition to high-definition television took place simultaneously and coincidentally with the transition to flat-panel displays (FPDs). On the Friday and Saturday after Thanksgiving 2006, American consumers bought, for the first time, more FPD televisions than CRT receivers. The proliferation of inexpensive, high-resolution FPDs capable of video-scanning rates provided those desiring HDTV with the experience NHK's engineers envisioned in the 1970s. Beyond broadcast television applications, manufacturers in the computer and cell phone industries also used monochrome and then color liquid crystal displays (LCDs) to increase the portability and versatility of their product. In doing so, and in tandem with increases in the data processing power of computers and ICs, FPDs also did as much to kill television as digital broadcasting and the growth of the Internet.

THE DEATH OF THE CATHODE-RAY TUBE TELEVISION, 2001–2006

In March 2006 Thomson closed its Picture Tube Development Group in Lancaster, Pennsylvania. The site where RCA's workers had mass-produced CRTs for the first televisions after World War II; where its engineers had turned RCA Laboratories' shadow-mask CRT into the world's technology for watching color TV; and where its licensing staff trained the rest of the world's companies in the manufacture of color picture tubes, was turned into an industrial space for lease. Shortly afterwards, Sony closed its CRT technology centers in Pittsburgh and San Diego, and its factories for making picture tube components in the Americas. The United States' role in the technology that defined sixty years of television had been reduced to an icon in the TiVo logo.

What happened? Engineers, scientists, and their sponsors stretched the CRT to the limits of practical use, but that was insufficient for HDTV applications and the advantages of FPDs. Sony introduced the world's largest color picture-tube receiver, 45 inches in diagonal screen width, in 1988. Others at Thomson/RCA and Samsung followed Sony's success in making and marketing its flat-screen Wega displays; raised the contrast and brightness; deflected the electron beams to over 125 degrees to flatten the cabinet; and changed the screen ratio from 4:3 to 16:9 to match the HDTV standards. The sets weighed over 200 pounds, not including the floor stand. Every increase in screen dimensions increased the weight of the glass containing the vacuum that defined the system. CRTs matched or surpassed the image qualities of an FPD in most respects, but consumers chose FPDs because of their flatness and mobility.

Americans still bought CRT TV receivers—nearly 25 million in 1996—well into the 2000s, when Asian manufacturers made and assembled all the components using the cheapest work forces and looser environmental standards in China and India. For a company like Samsung, a tidy profit remained in selling flat-faced, wide-screen, 30-inch, CRT televisions on-line and at Walmart and Target stores for around $700 for most of 2006, half the price of plasma or LCD televisions of similar dimensions. At the same time, its flat-panel engineers and scientists drove the development and production of the largest FPDs. Its executives managed to cater to both ends of the middle-class market, milking the profit margins possible with both mature and maturing technologies. For 2007, however, display manufacturers announced that none would offer more than one CRT receiver in the United States. Consumers would have a far greater choice of the LCD, plasma, or projection displays made in South Korea, Japan, Taiwan, and China.

THE TRIUMPH OF FLAT PANEL DISPLAYS, 1985–2006

The large-screen triumphs and expanding appeal of FPDs had been brewing well before the turn of the millennium, and contributed to the death of television, far from its conception in Europe. This was not simply a result of Asian engineering skill and business acumen, but of different national commitments and international alliances. Japanese, Korean, and Taiwanese consortiums accelerated the evolution of FPDs, aided by an American alliance that supported the design and supply of processing equipment, techniques, and materials. While no histories exist yet detailing the politics behind the commercial rivalry and the technical progress in FPDs, the results of the efforts speak for themselves.

Why the supporting role for the United States? The Defense Advanced Research Projects Agency and the armed forces underwrote pathbreaking developments at scattered companies, especially in circuitry for large, high-definition, plasma displays and active-matrix LCDs, but no American company could afford their commercialization. With Zenith's sale to South Korea's LG Electronics in 1995, American consumer electronics represented no more than brand names.

That year South Koreans organized their second, larger, FPD collaboration involving the government, universities, and large corporations. For a nation investing for a higher standard of living and international respect, there was no alternative to innovation. The alliances followed Samsung president Jong Bae Kim's 1985 conclusion that LCDs would replace picture tubes for television displays. He hired Chan Soo Oh to guide the company to leadership in that field, marked in 1996 by a 22-inch LCD.

Sharp's LCD breakthrough in 1988 coincided with a government-sponsored collaboration among twelve Japanese companies to produce a 40-inch LCD by 1995. They failed to meet that objective, but the effort improved the brightness, color, contrast, and viewing angle of LCDs as well as fabrication techniques, giving Japan 80 percent of the active-matrix market. Over the next ten years, Japanese and South Korean companies vied for bragging rights in fabricating the largest displays from single sheets of glass. Each leap forward, toward dimensions only dreamed of twenty years before, marked the establishment of a next-generation factory costing from a half to three billion dollars to make the displays. The cost drove the formation of international alliances, including Sony and Samsung, and Philips and LG Electronics. Sharp demonstrated a 29-inch LCD in 1996.

Without requiring massive transistor-addressed arrays like LCDs, PDPs enjoyed significant advantages in scaling up. In 1990, two years after Sharp showed its 14-inch LCD television, Fujitsu's engineers offered a 31-inch plasma TV made brighter by depositing the phosphors on the backplate where they reflected, rather than transmitted, light toward the viewer. While Sharp began selling 10-inch LCD TVs in 1995 for $1,650, Fujitsu and Philips introduced a 42-inch PDP for $12,000 that drew on Japanese and American innovations to improve contrast and gray scale. At the extreme edge of CRT practicality, this size became the standard for large-screen monitors, responsible for 70 percent of PDP sales over the next ten years.

In September 2006 NBC Universal bought Panasonic's first 103-inch high-definition plasma display for its *Football Night in America* show. The manufacturer, Matsushita, literally one-upped its South Korean rivals Samsung and LG Electronics, which demonstrated 102-inch PDPs in the previous two years. Panasonic's marketers expected to sell up to 5,000 of the 475-pound monitors in the next year for $70,000 each, not including installation costs.

Despite plasma's advantages in size, color, and cost, however, the display industry invested $79 billion in LCD manufacturing between 1996 and 2005, compared to $10 billion for PDPs. The commitment began as the FCC approved the HDTV standards, where the higher resolutions possible in LCDs for a given screen size justified investments in improving its other qualities. Sharp unveiled a 108-inch LCD, fabricated from a single sheet of glass, at the Consumer Electronics Show in Las Vegas in January 2007, making a quantum leap past its $11,000 65-inch LCD and topping the plasma competition in the process. By the end of 2006, reviewers of the technologies found them equivalent in quality while consumers bought more LCDs than PDP TVs. The price of the latter at 42 inches dropped $700 from 2005 to $1,300, while LCDs came down more sharply, from $3,000 to $1,600.

Flat-panel innovators not only ended an era of traditional TV displays but opened new venues for video. These extended beyond laptop and desktop computers, cell phones, and personal digital assistants (PDAs). In November 2006 Kopin announced that it had installed an LCD with a .16-inch screen and 300 × 225 resolution in a head-mounted display (HMD). This was not an unusual military application, as European and American armed forces had been buying HMDs for training simulators using digitized graphics since the 1970s. But Kopin sold its high-tech television system for children operating a wireless, $180 "Spy Video Car" and its infrared camera, just in time for the holiday shopping season.

THE DIFFUSION OF DIGITAL VIDEO, 1972–2006

The digitization of broadcast television meant that it became one more video option for users of all those displays. While the broadcast industry struggled with the switch to a digital system, computer scientists, engineers, enthusiasts, investors, entrepreneurs, and consumers poured their intellectual, financial, and creative resources into other digital video technologies. The oldest of these was video games, which Magnavox first innovated with its Odyssey system in 1972. Players interacted with challenges generated by a program in a computer that connected to a television via channel 3 or 4. Over thirty-five years, game designers developed increasingly sophisticated video graphics played from consoles, or Web sites, through a TV or computer display. The techniques attained full video frame rates in 1984 with RCA's announcement of digital-video, interactive, compact discs to be used with personal computers.

While the content and consumer electronics industries struggled with digital TV during the 1990s, consumers and entrepreneurs took matters literally into their own hands on the Internet and cellphone networks. University of Cambridge students mounted the first webcam, which watched a coffee pot in 1991, even before the invention of the World Wide Web, and it remained online for ten years. Junior Jennifer Ringley of Dickinson

College in Pennsylvania pioneered the online documentation of her life for seven years between 1996 and 2003, upgrading her JenniCam from a single monochrome camera to four color videocams on a Web site that received millions of visitors. Companies and governments installed video conferencing systems and millions of people installed their own webcams, the frame rate rising in inverse proportion to the price of cameras and broadband internet service, to enable true television over the internet. Victoria's Secret staged a 14-minute streaming webcast of its lingerie fashion show in 1999, attracting 1.5 million people to its Web site to watch the "jaggy and barely watchable" video ("1.5 Million").

Another informal television system arose out of the addition of video capability to digital cameras and cell phones, along with traditional video cameras. Examples of private citizens using video recording to check the abuse of authority received their first wide coverage in 1992. George Holliday used his brand-new camcorder to capture members of the Los Angeles Police Department beating Rodney King and delivered it to a local TV station. The ability to store and transfer increasing amounts of video data as electronic files ever more cheaply, and upload them to Internet sites for anyone to watch, removed the mediation role of broadcasters. In December 2006 an onlooker subverted the Iraqi government's attempt to sanitize Saddam Hussein's execution. His cellphone video file uploaded to YouTube, revealed the sectarian prejudices of the executioners and the tyrant's strength of character. Others downloaded it and edited in 1960s fashion footage and dance music, and posted that version online. People had turned cellphones into virtual TV stations, recording video and playing it back or uploading it, although the U.S. lagged behind Italy and Asian countries in offering wireless third-generation (3G) video sharing. While Japanese used 17,000,000 minutes sending video files in March 2005, television over the U.S. cellphone network awaited more corporate commitment, more bandwidth on the spectrum, and two-camera phones.

The number and variety of websites reflect the entire human experience more fully than ever before, or at least that of a significant population with access to the digital tools necessary to express their interests or their selves. Although Yahoo and AOL and other sites served as gatekeepers recommending or suppressing content, digital internetworks had the effect of removing technical, commercial, or cultural guardians from the medium. If anyone could find or make whatever he or she wanted on the Internet, then the mass media was close to transparency, and therefore in danger of losing its definition as a medium. Whether this constituted an improvement over television at any other stage in its life was moot. Anyone could select online evidence to suit their argument about the role of mass media in society, the value of popular access to production tools, and as the effects of content on viewers.

CONCLUSION: LOOKING FORWARD FROM 2006

In the fall of 2005, engineers from Japan's broadcast network, NHK, set up a demonstration of the next generation of digital television. Using a special 8-megabyte CCD camera, they captured footage containing 7680 × 4320 pixels in a frame, 60 frames per second, totaling about 24 gigabytes per second of data. In all it represented an image resolution sixteen times that of the highest HDTV standards. This quantity they divided and transmitted over 16 fiber-optic cables to be reassembled via a complex multiplexing technique at the World Exposition in Aichi, 161 miles away. "Super Hi-Vision has huge information and was difficult to transmit," said NHK senior engineer Mikio Maeda, but the network had shown what the world could expect to appreciate twenty years later (Hara, 2005, "Japan Demonstrates").

At the low-definition end of the video technology, business, and culture, nonbroadcast entrepreneurs held different visions for the promotion, development, and sale of video technologies over the informal networks created by people with common interests on the Internet and cell phone systems. All of these traded high-definition for the convenience of mobile video. They jostled for authority and investment in the emerging industry through the establishment of rival conferences. The Technology Marketing Corporation (TMC) built on eight years of publishing and trade show experience in the Internet telecommunications industry to offer its ITEXPO on Voice (and Video) over Internet Protocols (VOIP) in Fort Lauderdale, Florida, in January 2007. Tracy Swedlow, founder of the first interactive-TV newsletter in 1998, promoted "The TV of Tomorrow Show" in San Francisco in March. Jeff Pulver organized the second Video on the Net (VON) conference featuring "voice, video & vision" in San Jose the same month, while the first Internet TV Conference and Expo was scheduled for June in New York City.

In September 2006, at a conference of display executives and technologists at Kent State University in Ohio, Samsung's executive vice president for LCDs, Jun Hyung Souk, gave the keynote speech. His and other Asian companies were finally reaping the benefits from tens of billions of dollars invested in plasma and LCD factories. Samsung's eighth generation LCD fab, or factory, would be operational in October 2007, capable of manufacturing 3.6 million 52-inch LCDs a year. Souk forecast only six years into the future of video displays before admitting uncertainty, and with good reason. Everywhere he looked, sponsors of new display technologies, applications, and businesses sought the public's eyeballs, minds, and pocketbooks.

It was entirely possible, for example, that innovators in organic light-emitting diodes (OLEDs) would solve the technology's sensitivity to humidity and oxygen and overtake the FPD industry with cheaper, scalable, flexible, and therefore more versatile displays. The most visible video application for the low-power semiconductors had been for billboards in Shibuya Square in Tokyo and Times Square in New York City. In Ewing, New Jersey, and Cambridge, England, however, young companies promoted the flexibility and robustness of their monochrome, cardboard-thin, OLED displays on YouTube. One provided images a little over an inch wide showing a mountain biker in action. The other was a 10-inch, active-matrix display intended to replace paper with a softer ware, fabricated through inkjet printing, with all that success implied for the display and publishing industries. Beyond paper, Philips Research offered its flexible Lumalive OLED textiles at a European trade show in August for integration into high-fashion clothing. Just as engineers and scientists had developed technologies for applications unattainable by bulky and heavy CRTs, Philips's staff led other groups in devising one for uses unattainable by traditional FPDs.

Closer to Korea, in 2002 Sharp Electronics organized another Japanese consortium to innovate three-dimensional LCDs. Two years later it introduced the first laptops and displays to offer 3D without glasses, by adding a layer of parallax lines to the display. Turned on, they blocked columns of pixels from the left and right eyes of the viewer to give slightly different perspectives that the brain fused. The effect was dramatic as people "instinctively reached out to touch the images that appeared to float in space in front of the monitor" (Poor). Sharp promoted the first two generations of the technology's products to niche markets of scientific, graphics, and gaming consumers, but the early announcement of software to convert standard DVDs hinted at where Sharp's marketing staff might uncover a larger market.

Nonetheless, without knowing what technologies would prove popular in the next generation, Samsung's Souk was willing to imagine fulfillment of the display technologies in Philip K. Dick's 1956 short story *Minority Report*, as rendered by director Steven Spielberg in 2002. There, actor Tom Cruise watches holographic recordings of his family in 2054; runs down corridors surrounded by commercials; and manipulates video data displays by "conducting" them with his hands to assemble and synthesize the information he needs. Dick's dystopia did not rival the bleakness of

Larry and Andy Wachowski's *Matrix* films, in which a small group tries to break free of the digital virtual reality that tyrannical computers imposed on most of the world. Still, the cultural pessimism about the next generation of video technology mirrored darkly the sunnier visions of futurists from the 1870s to the 1940s, who had looked forward to the promise of television. Given Intel Corporation's goal of teraflop microprocessors by 2015 and the progressive fusion of imagery sources, it was not difficult to imagine people immersed in interactive landscapes fused from multiplying imagers, generated by graphic designers, or improvised from parts of reality and its virtual alternatives.

No one doubted that for the sake of diversion, surveillance, education, or productivity, corporations, governments, and individuals would continue to demand and pay for easier manipulation of rising oceans of digital data. The 2007 Consumer Electronics Show marked a watershed in coming to terms with the digital future of video and other media by the broadcast, computer, consumer electronics, and film industries. All realized they stood to benefit by helping connect broadcast and Internet content on the same household displays. CBS president Leslie Moonves proclaimed that "The days of old media and new media are over," but the public would not realize that until it embraced new names for devices like those formerly known as cell phones (Graham and Kessler, 2007).

In one of the show's opening addresses, Microsoft Corporation founder Bill Gates touted the "media home" to 140,000 attendees in Las Vegas, its digital media technologies intra- and interconnected through his company's software and hardware. In highlighting and catering to popular desires for "connected experiences" Gates channeled his visionary ancestor. David Sarnoff predicted such households in 1965: "A single integrated system means that the major channel of news, information, and entertainment in the home will combine all of the separate electronic instruments and printed means of communications today—television set, radio, newspaper, magazine, and book" (Sarnoff, 1968, 190).

Sony, Yahoo, Hewlett-Packard, DirecTV, and Apple joined Microsoft and Intel in demonstrating ways to watch Internet content on the large FPDs taking center stage in American homes. All knew that the resolution and frame rate of the imagery was embarrassingly poor, but that was secondary to the goal of merging the broadcast and Internet cultures in the home. Besides, they could anticipate its improvement with improved cameras, processors, and compression software. Alternately, Hollywood and the broadcast and cable stations arranged to deliver movies and programs to cell phones. Watching a film designed for a 40-foot screen on a 2-inch display—the same size that Ernst Alexanderson peered at in 1927—might not matter to "platform-agnostic" children. The film critic David Denby had to ask, however, "Where were movies going? Were they going any place good?" (Denby, 2007, 54) The questions applied to network television programs as well.

The consequences of television's death continue to play out wherever one lives, works, or plays. Inventors respond to new challenges in improving quality, size, and flexibility in video capture, processing, and display. Entrepreneurs exploit opportunities in the riot of new applications and the parallel problems of software and hardware compatibility. Manufacturers and governments wrestle over standards, spectrum, and the balance between pollution and the higher living standards generated by factories. Content producers struggle to maintain revenues across media platforms, and to limit copying and free distribution of content. Cultural pessimists and technological optimists use the worst and best of the Internet to justify their calls for action. Retailers seek out stylish products and new features with better profit margins, or cheaper goods and more customers. Through it all, however, consumers and users will make their prosaic choices on which technologies to adopt and engage, and which to ignore or abandon, in the finite time available in their waking hours, thus helping determine the life story of television's successors.

Glossary

Anode: A positively charged electrode that attracts electrons.

Bandwidth: The range of frequencies measured in cycles per second, or hertz, between given limits.

Binary: Related to a numeric system, base 2, that uses only 0 and 1.

Bit: A binary digit, 0 or 1.

Broadband: A transmission cable or wire capable of carrying a wide range of electromagnetic frequencies, measured at least in hundreds of kilohertz.

Byte: A string of eight bits processed by a computer.

Cathode: A negatively charged electrode that emits electrons.

C-band: A Federal Communications Commission (FCC) allocation of electromagnetic spectrum bandwidth for satellite communications: 5.925–6.425 GHz on the downlink to earth, and 3.7–4.2 GHz on the uplink to the satellite.

CCD: Charge-Coupled Device, a solid-state technology used in video and digital cameras.

CMOS: Complementary Metal-Oxide Semiconductor, a very low-power transistor circuit used in virtually all electronics and growing numbers of digital cameras.

CRT: Cathode-Ray Tube, usually a glass vacuum tube containing a cathode, whose electrons are focused in a beam toward a phosphor-coated faceplate to portray a signal.

Digital: Related to information, or data, that has been converted to or is in numeric, usually binary, form.

Digitize: To convert information—from audio and video signals, for example—into numbers, usually on a binary system of counting.

Electrode: An electronic element that emits, controls, or attracts electrons.

Electromagnetic: Relating to the generation of magnetic fields by electricity or vice versa; or, the range of radiation in the environment extending from very long radio waves through visible light to gamma rays.

Electron: Negatively charged subatomic particles that orbit each atom's nucleus at the speed of light.

Electron gun: Device in the neck of a picture or camera tube whose cathodes emit, control, and focus a beam of electrons.

Electrostatic: relating to attraction or repulsion by electric charge.

ENG: Electronic news-gathering, with reference to lightweight portable broadcast television cameras and equipment.

Field: A set of scanning lines that comprise part or all of a television picture. In interlaced systems, two or more fields make up a complete frame of video.

FPD: Flat-panel display.

Frame: In television, a complete image transmitted as part of a series.

Geosynchronous orbit: Path around the earth that matches the earth's rotation, in which the orbiting object appears to be stationary when viewed from the earth.

Gigahertz (GHz): Billions of cycles per second, or hertz, in an electromagnetic wave.

HD-MAC: High-Definition Multiplexed Analog Components, the European high-definition standard from 1986–1992.

Head end: The reception facilities of a local cable operator, which can receive signals by broadcast, microwave, or satellite antennas and then distribute them to wired customers.

Hertz: International standard term for cycles per second, named after Heinrich Hertz.

Iconoscope: RCA-trademarked term from the Greek words for image and sight, used for the company's first television camera tube.

ICs: Integrated Circuits, collections of active and passive electronic components—transistors, resistors, and capacitors—that can be etched and deposited on a substrate.

Interlaced scanning: In which alternate lines of an image are scanned or captured for transmission, and recombined at the display for each frame of video.

K_a-band: A Federal Communications Commission (FCC) allocation of electromagnetic spectrum bandwidth for satellite communications: 18.3–18.8 GHz on the downlink to earth, and 19.7–20.2 GHz on the uplink to the satellite.

K_u-band: A Federal Communications Commission (FCC) allocation of electromagnetic spectrum bandwidth for satellite communications: 14–14.5 GHz on the downlink to earth, and 11.7–12.2 GHz on the uplink to the satellite.

Kinescope: RCA-trademarked term from the Greek words for "motion" and "sight," adopted by the television industry first for all cathode-ray tubes for television displays, and second for films recording programs off the face of such a tube.

LCD: Liquid Crystal Display.

LED: Light-Emitting Diode.

Megabyte (MB): Millions of bytes: there are 8 binary bits to a byte.

Megahertz (MHz): Millions of cycles per second, or hertz, in an electromagnetic wave.

Modulate: To change or vary an electronic signal's wavelength, amplitude, or phase.

Monochrome: Single-color video signal, usually black and white but determined by the wavelength of light emitted in a display.

MPEG-2: An international standard for the digital coding and compression of broadcast video and related audio signals and DVDs at a rate of 50 to 1.

MPEG-4: An international standard for the digital coding and compression of video and related audio signals at more efficient compression rates than MPEG-2, for broadcast and Internet media.

Orthicon: RCA's second camera tube, from Greek words for "straight" and "image."

Packetization: The process of breaking a digital bitstream of data in discrete pieces and assigning address and content information to them.

PDP: Plasma Display Panel.

Photocell: A device that converts a light signal into electricity.

Photoconductive: Related to electronic conductivity within a material under exposure to light.

Photoelectric: Related to electronic effects stimulated by exposure to light radiation.

Photoemissive: The property of a material to emit electrons when exposed to light

Photon: Particle of light.

Pixel: A single point of light collection or illumination on a digital camera or display.

Primary colors: In television, the three colors red, green, and blue, that are emitted or reflected from a subject and can be combined, or added, to generate the entire color spectrum.

Progressive scanning: In which each line of an image is scanned in succession for a camera or display; used in Nipkow disc televisions, CRT computer monitors, and certain flat-panel displays.

Raster: The number of lines that constitute one frame of video on a display.

Scanning: Capture of light values in an image by dividing it into any given number of elements for an equivalent frame.

Semiconductor: Solid material whose ability to conduct electrons can be controlled for various purposes.

TFT: Thin-Film Transistor, built on an insulator, like glass.

Transistor: Semiconducting device containing three electrodes.

Transponder: Transmitter-responder device that receives a signal at one frequency, amplifies it, and sends it at a different frequency.

Bibliography

Books like this one rely on the work of writers who have done their research in archival collections of letters, notebooks, reports, photographs; in the primary technical literature; in the oral histories and memoirs of participants; and in Web sites and Web pages that contain all of the above. Web references that no longer appear at the listed URLs might be found by entering the URL in the Wayback Machine at the Internet Archives.

"About CBN." *Christian Broadcasting Network* (cbn.com/about/), visited June 26, 2005.

Abramson, Albert. *The History of Television, 1880 to 1941.* Jefferson, NC: McFarland, 1987.

———. *Zworykin: Pioneer of Television.* Urbana, IL: University of Illinois Press, 1995.

———. *The History of Television, 1942 to 2000.* Jefferson, NC: McFarland, 2003.

Bannister, Jennifer Burton. "From Laboratory to Living Room: The Development of Television in the United States, 1920–1960." Ph.D. dissertation, Carnegie Mellon University, 2001.

Barnouw, Erik. *Image Empire: A History of Broadcasting in the United States, Vol. III—From 1953.* New York: Oxford University Press, 1970.

———. *Tube of Plenty: The Evolution of American Television*, 2nd rev. ed. New York: Oxford University Press, 1990.

Bedell Smith, Sally. *In All His Glory: The Life of William S. Paley.* New York: Simon and Schuster, 1990.

Bilby, Kenneth. *The General: David Sarnoff and the Rise of the Communications Industry.* New York: Harper & Row, 1986.

Breddels, Paul, and Walter W. Zywottek. "Merck Investors Conference Call, November 29, 2006" (merck.de/servlet/PB/show/1566670/Conference%20Call_Bear%20Stearns_29.11.2006.pdf), visited January 6, 2007.

Brinkley, Joel. *Defining Vision: The Battle for the Future of Television.* New York: Harcourt Brace, 1997.

Brittain, James E. *Alexanderson: Pioneer in American Electrical Engineering.* Baltimore, MD: Johns Hopkins University Press, 1992.

Brown, George. *And Part of Which I Was: Recollections of a Research Engineer.* Princeton, NJ: Angus Cupar, 1982.

Bucher, Elmer E. *Television and David Sarnoff,* Part 12. Bound typescript at David Sarnoff Library, Princeton, NJ, 1952.

Burns, Russell. W. *Television: An International History of the Formative Years.* London: The Institution of Electrical Engineers, 1998.

———. *John Logie Baird, Television Pioneer.* London: The Institution of Electrical Engineers, 2000.

Carmack, Carmen. "Mobile Video Phones, For Real." *PC Today*, 4(9) (September 2006) (pctoday.com/editorial/article.asp?article=articles/2006/t0409/04t09/04t09.asp&guid=), visited December 20, 2006.

Castellano, Joseph A. *Liquid Gold: The Story of Liquid Crystal Displays and the Creation of an Industry.* Singapore: World Scientific Publishing Company, 2005.

———. "Modifying Light." *American Scientist*, 94(5) (September–October 2006).

Chan, Rodney. "Flat-panel TVs See Spectacular Black Friday Sales, NPD says." *DigiTimes.com*, December 5, 2006 (digitimes.com/displays/a20061205PR206.html), visited December 8, 2006.

Chandler, Alfred D., Jr. *Inventing the Electronic Century: The Epic Story of the Consumer Electronics and Computer Industries.* New York: Free Press, 2001.

"China Lake Weapons Digest: 50 Years of Providing the Fleet with the Tools of the Trade" [originally written for *The Rocketeer*, China Lake Naval Air Warfare Center, November 4, 1993] (nawcwpns.navy.mil/clmf/weapdig.html), visited April 3, 2006.

Cooper, Bob. "Television: The Technology That Changed Our Lives." *Early Television Foundation* (earlytelevision.org/color_tv_cooper.html), visited June 18, 2006.

———. "In Perspective 2 (Updated content 20 June 2006)" (bobcooper.tv/inperspective2.htm), visited June 30, 2006.

Cooper, Tom. "Headless Fighters: USAF Reconnaissance-UAVs over Vietnam." *Indochina Database* (acig.org/artman/publish/article_344.shtml), visited April 3, 2006.

Crane, Rhonda J. *The Politics of International Standards: France and the Color TV War.* Norwood, NJ: Ablex Publishing Corporation, 1979.

Denby, David. "Big Pictures." *The New Yorker,* January 8, 2007.

Diehl, Richard. "The Time Line of Extinct Video Tape Recorders." *LabGuy's World* (labguysworld.com/VTR_TimeLine.htm), visited June 26, 2006.

"East Pushes the Way with Internet Protocol TV." *CNN.com Digital Biz* (cnn.com/2006/TECH/biztech/11/26/iptv/index.html), visited December 16, 2006.

Everson, George. *The Story of Television*. New York: W. W. Norton & Company, 1949.

Fahrney, Delmar S. "The Birth of Guided Missiles." *Naval Institute Proceedings*, December 1980.

Farley, Tom. "The Cell-Phone Revolution." *American Heritage of Invention & Technology*, 22(3), Winter 2007.

Fink, Donald. "The Forces at Work Behind the NTSC Standards." A paper presented at the *122nd Annual SMPTE Technical Conference*, November 9–14, 1980, New York. Reproduced at *Williams Labs: Electronics Tutorials* (ntsc-tv.com/ntsc-main-01.htm#Monochrome), visited July 15, 2006.

"First Commercial Installation of World's Largest 103 inch PDP." *Gizmag.com*, September 1, 2006 (gizmag.com/go/6085/), visited December 8, 2006.

Fisher, David E., and Marshall Fisher. *Tube: The Invention of Television*. Washington, DC: Counterpoint, 1996.

Flory, Rob. "Broadcast, Scientific, and Industrial Television." *Les Flory Television and Electronics Page* (home.earthlink.net/~robandpj/id10.html), visited June 26, 2006.

"FOLED." *YouTube* (youtube.com/watch?v=jCkpnwRIQ6w&eurl=), visited December 19, 2006.

Glass, Katalin Tihanyi. "The Iconoscope: Kalman Tihanyi and the Development of Modern Television" (mtesz.hu/scitech/history/tihanyi/index.html), visited May 4, 1999, published earlier as "The Iconoscope and the Development of Modern Television." *Technikatorteneti Szemle* [journal of the Hungarian Museum of Technology], XI (1993), 173–199.

Glover, Daniel R. "NASA Experimental Communications Satellites, 1958–1995." *NASA History Division* (history.nasa.gov/SP-4217/ch6.htm), visited June 25, 2006.

Godfrey, Donald. *Philo T. Farnsworth: The Father of Television*. Salt Lake City: University of Utah Press, 2001.

Goebel, Greg. "Modern Glide Bombs." *Dumb Bombs & Smart Munitions* (vectorsite.net/twbomb5.html), visited April 9, 2006.

Graham, Jefferson, and Michelle Kessler. "Video Leads Parade As Old Media and New Media Hook Up." *USA Today* (usatoday.com/tech/products/2007-01-07-ces-media_x.htm?POE=TECISVA), visited January 7, 2007.

Graves, George. "Local HDTV by Antenna." *HDTV Solutions* (hdtvsolutions.com/HDTV_antennas.htm), visited December 16, 2006.

Günthör, Frank. *Frank's Handheld TV Pages:* Part 5, "The Seiko TV-Watch;" Part 6, "The Short 'History' of Pocket-TV" (taschenfernseher.de/index-e.htm), visited November 19, 2006.

Hara, Yoshiko. "Japan Demonstrates Next-gen TV Broadcast." *EE Times*, November 3, 2005 (eetimes.com/showArticle.jhtml?articleID=173402762), visited December 15, 2006.

———. "Matsushita to Offer 103-inch PDP." *Video/Imaging Design Line*, courtesy *EE News*, July 21, 2006 (videsignline.com/products/191000900), visited December 8, 2006.

Harris, Tom. "How Plasma Displays Work." *HowStuffWorks* (electronics. howstuffworks.com/plasma-display1.htm), visited December 10, 2006.

Hilmes, Michele, ed. *The Television History Book*. London: BFI Publishing, 2003.

The History of Television: A Brief History of TV Technology in Japan (nhk.or.jp/strl/aboutstrl/evolution-of-tv-en/index-e.htm), visited June 18, 2006.

Hornbeck, Larry J. "From Cathode Rays to Digital Micromirrors: A History of Electronic Projection Display Technology." *TI Technical Journal,* July–September 1998 (focus.ti.com/download/dlpdmd/166_History_Electronic_Proj_Tech_Hornbeck.pdf), visited December 2, 2006.

Hutchinson, Jamie. "Plasma Display Panels: The Colorful History of an Illinois Technology." *Electrical and Computer Engineering Alumni News*, XXXVI(1) (University of Illinois at Urbana-Champaign, Winter 2002-03, on-line at ece.uiuc.edu/alumni/w02-03/plasma_history.html), visited December 10, 2006.

Iams, Harley. "An Outline of Iconoscope Development between April, 1930 and October, 1933." Camden, NJ: RCA Victor Company, October 23, 1933, at "Iams Iconoscope 1933." *David Sarnoff Library* (davidsarnoff.org/gallery-iams/gallery-ia.htm), visited April 3, 2006.

Inglis, Andrew F. *Behind the Tube: A History of Broadcasting Technology and Business.* Boston, MA: Focal Press, 1990.

International Conference on the History of Television—From the Early Days to the Present, 13–15 1986. IEE conference publication 271. London: Institution of Electrical Engineers, 1986.

Jacobson, Sava. "CBS and Color Television, 1949–1951." *David Sarnoff Library* (davidsarnoff.org/jac.htm), visited May 27, 2006.

Johnstone, Bob. *We Were Burning: Japanese Entrepreneurs and the Forging of the Electronic Age.* New York: Basic Books/Westview Press, 1999.

"Kalman Tihanyi (1897–1947), Automatic Sighting and Directing Devices for Torpedos, Guns and Other Apparatus (1929–34)." In *Aviation Pioneers: An Anthology* (ctie.monash.edu.au/hargrave/tihanyi.html), visited April 3, 2006.

"Kalman Tihanyi's 1926 Patent Application 'Radioskop' (Hungary)." *UNESCO Memory of the World Register—Nomination Form* (unesco.org/webworld/mdm/2001/eng/hungary/tihanyi/form.html), visited April 3, 2006.

Kawamoto, Hirohisa. "The History of Liquid Crystal Displays." *Proceedings of the IEEE*, 90(4) (April 2002), 460–500.

Keating, Jeffrey, and Mark Long. *The World of Satellite Television.* Summertown, TN: The Book Publishing Company, 1983.

Kisseloff, Jeff. *The Box: An Oral History of Television, 1920–1961.* New York: Viking, 1995.

"Kopin's Microdisplay Incorporated into Wild Planet's Spy Video Car^{TM}," www.elis.ugent.be/ELISgroups/tfcg/microdis/kopinspyvideocar.pdf, visited December 15, 2006.

Kornblum, Janet. "With Tiny Tech, a Video Little Brother Is Always Watching." *USA Today* (usatoday.com/tech/news/2006-12-05-little-brother-cell-video_x.htm), visited December 19, 2006.

Krichoveev, Mark I. *The First Twenty Years of HDTV: 1972–1992.* White Plains, NY: Society of Motion Picture and Television Engineers, 1993.

Lardner, James. *Fast Forward: Hollywood, the Japanese, and the Onslaught of the VCR.* New York and London: W. W. Norton & Company, 1987.

Lee, Chung-Shing, and Michael Pecht. "Flat Panel Displays: What's Going on in East Asia Outside Japan" (calce.umd.edu/general/AsianElectronics/Articles/DISPLAY.htm), visited December 30, 2006.

Lemmon, Sumner. "Samsung Says LCDs Will Take on Plasma in Market for 50-inch TVs." *PC World* (pcworld.com/article/id,127854-c,tv/article.html), visited December 16, 2006.

Lubar, Steven. *InfoCulture: The Smithsonian Book of Information Age Inventions.* Boston, MA: Houghton Mifflin, 1993.

Lyons, Eugene. *David Sarnoff.* New York: Harper & Row, 1966.

Lyons, Nick. *The Sony Vision.* New York: Crown, 1976.

Maclaurin, W. Rupert. *Invention and Innovation in the Radio Industry.* New York: The Macmillan Company, 1949.

"Magnavox Odyssey: First Home Video Console" (pong-story.com/odyssey.htm#dating), visited January 6, 2007.

McBride, Sarah. "TV + the Web = ?" *The Journal Report, Wall Street Journal*, May 15, 2006.

McLean, Donald F. "'Phonovision': The World's First Recordings of Television." *The World's Earliest Television Recordings* (tvdawn.com/tv1strx.htm), visited June 16, 2006.

Montalbano, Elizabeth. "Gates Touts Digital Home." *PC World* (pcworld.com/article/id,128436-c,tradeshows/article.html), visited January 7, 2007.

"Motorola Debuts First Ever Nano Emissive Flat Screen Display Prototype." *PhysOrg.com* (physorg.com/news4031.html), visited December 17, 2006.

Negroponte, Nicholas. *Being Digital.* New York: Knopf, 1995; see also *An OBS Cyberspace Extension of Being Digital* (archives.obs-us.com/obs/english/books/nn/bdintro.htmJ), visited December 30, 2006.

Nist, Ken. "An HDTV Primer." *HDTV Magazine* (hdtvprimer.com/#LineDoubler), visited December 14, 2006.

O'Leary, Michael. "Incredible Cat." *Air Classics*, January 2000 (findarticles.com/p/articles/mi_qa3901/is_200001/ai_n8901621), visited April 3, 2006.

"1.5 Million People Tune into Victoria's Secret's Fashion Show." *Geek.com* (geek.com/techupdate/feb99/secretstats.htm), visited January 6, 2007.

Parsch, Andreas. "Martin Marietta AGM-62 Walleye" (designation-systems.net/dusrm/m-62.html), visited April 3, 2006.

Perry, Tekla S. "Among the Classics: The VCR, 1975." *IEEE Spectrum*, 25(11) (1988), 112–115.

"Philips Research Shows Off Technology That Displays Electronic Images On Clothes." *ScienceDaily*, August 30, 2006 (sciencedaily.com/releases/2006/08/060828085919.htm), visited December 16, 2006.

"Plasma Loses Ground to LCD TVs." *CNN.com Digital Biz*, November 30, 2006 (cnn.com/2006/TECH/biztech/11/26/lcd.plasma.reut/index.html), visited December 15, 2006.

"Plastic Logic to Show High Resolution Flexible Active-matrix Display at Plastic Electronics 2006." *Plastic Logic* (plasticlogic.com/news-detail.php?id=295), visited December 19, 2006.

Reitan, Edward Howard, Jr. *The Following Program Is Brought to You in Living Color* (novia.net/~ereitan/), visited December 2, 2006.

Rhea, Henry E. "Airborne Television Is an RCA Development." *Broadcast News*, 43 (June 1946).

Ritchie, Michael. *Please Stand By: A Prehistory of Television*. Woodstock, NY: The Overlook Press, 1994.

Roehl, Richard, and Hal R. Varian. *Circulating Libraries and Video Rental Stores* (sims.berkeley.edu/~hal/Papers/history/), visited June 29, 2006.

Sarnoff, David. *Looking Ahead: The Papers of David Sarnoff.* New York: McGraw-Hill Book Company, 1968.

Schreiber, William F. *William F. Schreiber* (wfschreiber.org/pageone.html), visited December 16, 2006.

Sharp Electronics. "3D Display Technology." *Ingenuity in 3D* (sharp3d.com/technology/), visited December 19, 2006.

Sipress, Alan. "FCC Vote a Victory for Phone Companies." *WashingtonPost.com*, December 21, 2006 (washingtonpost.com/wp-dyn/content/article/2006/12/20/AR2006122000779.html), visited December 30, 2006.

Skilos, Richard. "The Hat Trick That Didn't Happen." *New York Times*, December 10, 2006 (nytimes.com/2006/12/10/business/yourmoney/10frenzy.html?ex=1166590800&en=c35401fccafc2a4d&ei=5070), visited December 10, 2006.

Slotten, Hugh R. *Radio and Television Regulation: Broadcast Technology in the United States, 1920–1960*. Baltimore, MD: The Johns Hopkins University Press, 1998.

Smith, Sally Bedell. *In All His Glory: The Life of William S. Paley, the Legendary Tycoon and His Brilliant Circle*. New York: Simon and Schuster, 1990.

Sony History, Chapters 8, 10, 12, 14 (www.sony.net/Fun/SH/), visited June 25, 2006.

Stashower, Daniel. *The Boy Genius and the Mogul: The Untold Story of Television*. New York: Broadway, 2002.

Sterling, Christopher H., and John Michael Kittross. *Stay Tuned: A History of American Broadcasting*, 3rd ed. Mahwah, NJ: Lawrence Erlbaum Associates, 2002.

Strover, Sharon. "United States: Cable Television." *The Museum of Broadcast Communications* (museum.tv/archives/etv/U/htmlU/unitedstatesc/unitedstatesc. htm), visited June 24, 2006.

Susumu Yoshida, Electrical Engineer. An oral history conducted by William Aspray. IEEE History Center, New Brunswick, NJ, 1994 (ieee.org/portal/cms_ docs_iportals/iportals/aboutus/history_center/oral_history/pdfs/Yoshida208. pdf), visited December 8, 2006.

"Swinging Saddam." *YouTube* (youtube.com/watch?v=DzzDKP2z59E), visited December 30, 2006.

Tae-gyu, Kim. "LG.Philips Develops 100-Inch LCD." *Korea Times*, March 8, 2006 (http://times.hankooki.com/lpage/tech/200603/kt2006030818081211780. htm), visited December 2, 2006.

Udelson, Joseph. *The Great Television Race: A History of American Television Industry 1925–1941.* Birmingham, AL: The University of Alabama Press, 1982.

Unger, Jason. "The Future of HD: Not So Bright?" *CE Pro*, May 24, 2006 (cepro.com/news/editorial/11873.html), visited December 16, 2006.

Von Schilling, James. *The Magic Window: American Television, 1939–1953.* New York: The Haworth Press, 2003.

Watters, Darren. "Bill Gates Hails 'digital decade'." *BBC News* (news.bbc.co.uk/1/ hi/technology/6239975.stm), visited January 7, 2007.

Webb, Richard C. *Eye of the Peacock: A Story About the Engineers Who Brought You Television.* New York: Wiley Press, 2005.

Weber, Larry F. "History of the Plasma Display Panel." *IEEE Transactions on Plasma Science*, 34(2) (April 2006).

Weimer, Paul K. "An Interview Conducted by Mark Heyer and Al Pinsky." *IEEE History Center*, July 8, 1975 (ieee.org/portal/cms_docs_iportals/iportals/ aboutus/history_center/oral_history/pdfs/Weimer022.pdf), visited June 26, 2006.

———. "A Historical Review of the Development of Television Pickup Devices (1930–1976)." *IEEE Transactions on Electron Devices*, 23(7) (July 1976).

Wikipedia: "Anik (Satellite)," "Cable Television," "Channel Surfing," "Communications Satellite," "Couch Potato," "List of Communications Satellite Firsts," "Orbita" (en.wikipedia.org/wiki/), visited June 24–26, 2006, and "Jennicam," visited December 20, 2006.

Williams, Martyn. "Sharp Unwraps Behemoth 108-Inch LCD TV." *PC World* (pcworld.com/article/id,128427-page,1/article.html), visited January 8, 2007.

Wolpin, Stewart. "The Race to Video." *American Heritage of Invention & Technology*, 10(2) (Fall 1994), 52–58.

WZBN-25 New Jersey's Capital News Station (wzbntv25.com/index.html), visited December 31, 2006.

Zworykin, V. K. "Flying Torpedo with an Electric Eye." *RCA Review*, VI(3) (September 1946).

Index

About the Author

ALEXANDER B. MAGOUN is Executive Director of the David Sarnoff Library.